Collins Guide to Modern TECHNOLOGY

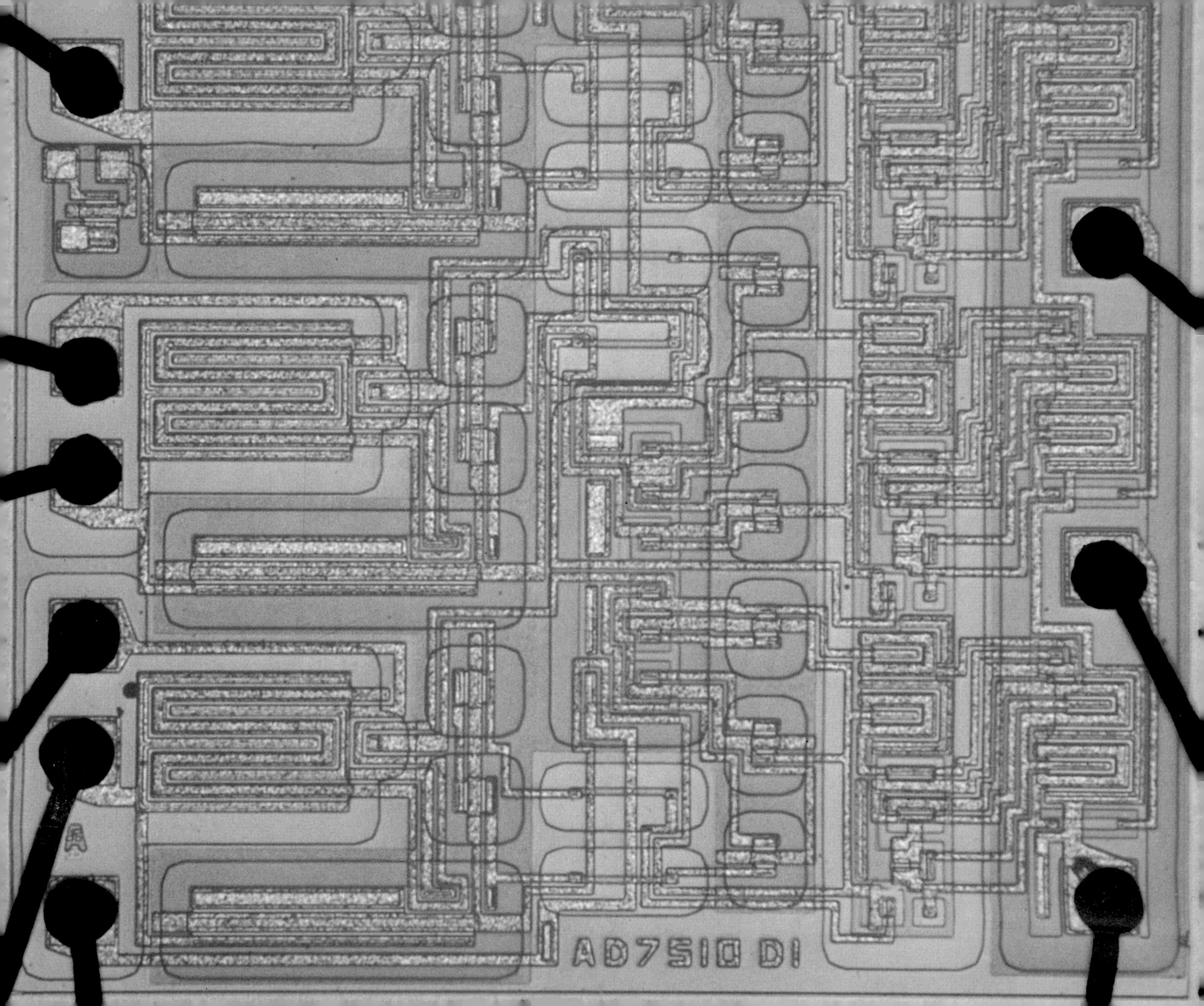
AD7510 DI

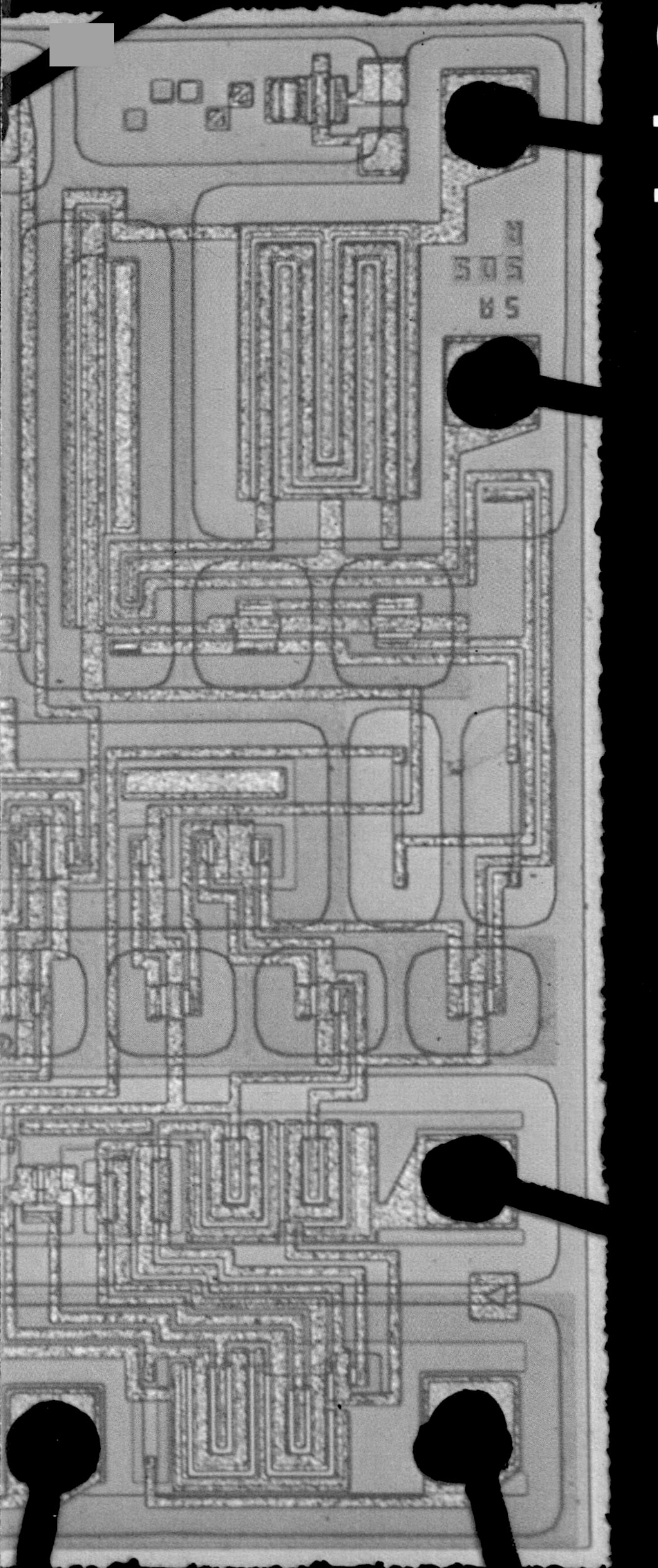

Collins Guide to Modern TECHNOLOGY

Robin Kerrod

Editorial Adviser: Professor Ray Wild,
Professor of Engineering and Management Systems,
Brunel University

Collins

Introduction

About two million years ago our primitive ape-like ancestors in Africa began to make tools by chipping flakes from pieces of flint. In so doing they took a gigantic leap forwards along the path to civilization. They had begun to invent things, a trait that set them apart from the other animals. They had begun to alter their environment for their own benefit. And ever since human beings have been improving their lives by making practical use of their discoveries and inventions – in other words, by technology.

The progress of technology was generally very slow until about two centuries ago. Then, in the 1700s, the pace began to quicken, and a series of inventions, first in textile-making, led to what is known as the Industrial Revolution. The Revolution was brought about by mechanization – the widespread use of machines – and the availability of a reliable power source to run them. This was the steam engine. The application of the steam engine to railways and ships began a revolution in transport. Instant, long-distance communications became possible following the invention of the telegraph and telephone. As invention bred invention civilization began advancing quickly along a broad front.

The Industrial Revolution laid the foundations of modern technology – the technology that has brought us offshore oil rigs and nuclear reactors; combine harvesters and automated milking parlours; synthetic fibres and industrial robots; suspension bridges and hydroelectric dams; turbo-charged racing cars and "bullet trains"; supersonic airliners and hovercraft; word processors and video recorders; pocket computers and sonar scanners; radio telescopes and space shuttles.

We concentrate in this book on the beneficial aspects of technology, but we must remember that technology has another face. Hands that forge ploughshares for the good of mankind can also forge swords for mankind's destruction. It has always been so. We now have what appears to be the ultimate weapon – the terrible hydrogen bomb. But perhaps even this is not all bad. The prospects are good that, with advancing technology, we shall be able to control the energy-producing process behind the bomb. And that could mean abundant power for the whole world for the foreseeable future.

History, it is often said, repeats itself. Chips of flint were in the forefront of the earliest technology. Today, closely related chips are in the forefront of the latest technology. They are silicon chips, made from quartz, which is chemically identical to flint. These chips, too, promise to transform human society, but in a few years, not two million.

Written and designed by Robin Kerrod
Illustrations and diagrams by Ian McIntosh, John and Mike Gilkes, Bryan Foster

First published 1983 by William Collins Sons & Co Ltd
London · Glasgow · Sydney · Auckland · Toronto · Johannesburg

ISBN 0 00 102134–6

Origination by Dot Gradations Ltd
Made and printed in Singapore by Toppan Printing Co (S) Pte Ltd

Contents

Chapter 1

Fuel and Power

Our technological world demands increasing supplies of fuel and power to turn the wheels of industry and transport and improve our living standards at home. Coal, oil and natural gas still provide the bulk of our energy needs. Oil products are burned in vast quantities in the engines of cars, trucks, planes and ships. All three fuels are burned in power stations to produce electricity, our most convenient form of energy. Other power stations use the nuclear "fuel" uranium.

However, coal, oil, natural gas and uranium are mined from the Earth's crust, and supplies of them will one day run out. Then we shall have to turn to alternative sources of energy if our civilization is to continue to flourish. We shall need to exploit the energy sources in nature on a vast scale – the rivers, tides, sunshine, wind and waves. Eventually we must find how to tap, through nuclear fusion, the very energy that powers the universe.

Boston at night, ablaze with light. The modern city, and modern civilization, could not exist without abundant supplies of fuel and power.

Coal

Coal was the world's most important fuel for hundreds of years. It was first mined in huge quantities in the 18th century at the time of the Industrial Revolution, when coal was burned to provide heat for steam-engine boilers. It was also made into coke for use in iron-smelting. For the home it was used to make coal gas for lighting and heating. In the late 1800s it was used in the new electricity power stations. By then coal had also become an invaluable source of organic chemicals for the chemical industry. These were made from COAL TAR.

Today things have changed. Coal provides less than a third of the world's energy. It. place as top fuel has been overtaken by oil, or petroleum. Little coal gas is now made because of plentiful supplies of natural gas. However, coal is still used in vast quantities to make coke for smelting and as a fuel in power stations. In some countries, notably South Africa, it is being used to make oil.

The United States and Russia are the world's largest coal producers. Output in the United States is some 800 million tonnes a year, nearly 30 per cent of the world's total output. China, Germany, Britain and Poland are also major coal producers. All these countries have vast reserves of coal, which should last for many centuries. This is fortunate because supplies of oil and natural gas could run out early next century. Then coal could once again become the dominant fuel.

Below: A line of hydraulically-operated props used in longwall mining. The props are moved forwards one by one after the shearer has passed.

Grades of Coal

Coal is a fossil fuel. It is the remains of huge plants that lived in tropical swamps hundreds of millions of years ago, mostly during the Carboniferous Period of the Earth's history. The plants died and decayed and gradually became covered by layers of mud and sand. Over the years heat and pressure in the Earth's crust changed the mud and sand into layers of solid rock, and the decayed plant remains into seams of coal.

Three main types of coal are found today, which differ in the amount of carbon they contain and in their heating value. The best-quality coal is anthracite, which is hard, shiny black and quite clean to the touch. Next comes bituminous coal, which is dirtier and contains less carbon but still has a high heating value. The lowest-grade coal is lignite, or brown coal. A low-grade fossil fuel that did not quite become coal is PEAT.

Opencasting

When coal lies on or near the surface, it can be extracted relatively easily by ordinary surface mining, often called opencast, or open-pit mining. The technique is simple. Any soil ("overburden") covering the coal seam is removed by excavators, and then power shovels load the exposed coal into rail cars or trucks. A common method is called

Above: A shearer operating in a longwall mine. It cuts coal from the coalface and deposits it onto the conveyor (bottom) for removal.

strip-mining. The coalfield is worked in a series of long strips. The overburden from each strip is used to fill in the trench where the coal has been excavated from a previous strip.

The excavators used in opencast operations are some of the biggest machines in the world. In the vast lignite mines in West Germany there are giant bucket-wheel excavators that can dig out 200,000 cubic metres of overburden a day. The biggest dragline excavators are at work in the United States. Biggest is the 12,000-tonne "Big Muskie" at the Muskingum mine in Ohio, whose excavating bucket is big enough to pick up a dumper truck!

Mining Underground

With modern excavators, surface mining may be possible down to a depth of 100–200 metres or more. Deeper coal seams have to be mined underground. Vertical shafts are sunk in the coalfield, and then horizontal tunnels are driven outwards from them into the coal seams. Extracting, or "winning" the coal is nowadays highly mechanized. Not only is the coal loaded and transported underground by machine, it is also cut by machine. This is possible because coal is relatively soft, unlike most other mined materials.

The most advanced method is called longwall mining because it takes place along a coalface hundreds of metres long. It is done by specialized power-loaders, such as shearers, which cut a slice off the face as they move forwards. The coal falls into a heavy chain conveyor, which removes it. A system of powered supports, or props, hold up the roof over the working area.

As the cutting proceeds and the coalface advances, the props are moved forwards. The roof behind is allowed to cave in. This system is highly productive. In 1982 a longwall mining team at the Sunnyside Mine in Utah produced 20,000 tonnes of coal in just 24 hours.

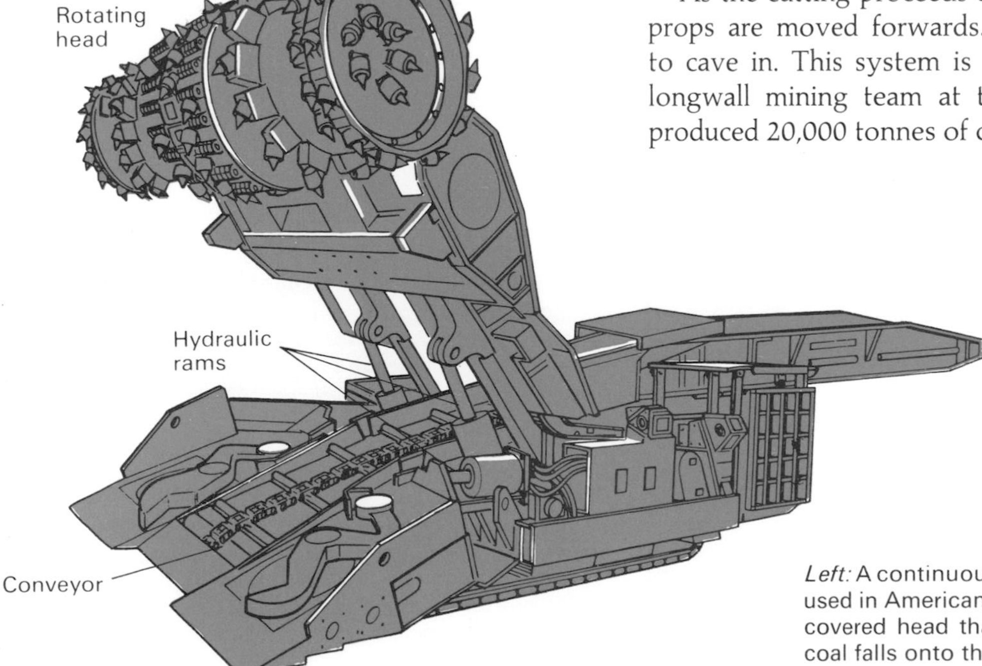

Left: A continuous-mining machine of a type widely used in American coal mines. It has a rotating stud-covered head that digs into the coalface. The cut coal falls onto the conveyor, which carries it to the rear and feeds it to a wagon, or another conveyor.

Oil and Gas 1

In 1859 in Pennsylvania in the United States, Edwin Drake launched a revolution by drilling the world's first oil well. Until then the world had relied almost entirely on coal for its fuel and power. Then gradually, oil began to take over.

At first, kerosene for lamps was the main oil product. Then when cars and aeroplanes came onto the scene petrol was in demand for their engines. Petrol is still the most sought-after product made from oil, or petroleum ("rock oil") as it is properly called. Kerosene is also still in demand, not for lamps, but as fuel for jet engines. Other fuels obtained from petroleum include diesel oil and heating oil.

However, petroleum is not just a fuel. It is also a reservoir of chemicals. It is made up of a mixture of substances called HYDROCARBONS, which are invaluable raw materials for the chemical industry. All these products, fuels and chemicals, are obtained by processing petroleum in an oil refinery (see page 56).

Above: The Umm Shaif offshore oil complex in Abu Dhabi, in the Persian Gulf. The complex includes not only production facilities, but also an accommodation platform for the operating crew. The Persian Gulf is one of the richest oil-bearing regions in the world.

Oil Production

Oil supplies nearly half of the energy consumed in the world today. In the early 1980s about 6400 million litres of oil were produced – and consumed – every day throughout the world. (Oil production is usually measured in terms of barrels of 160 litres, so daily production is about 40 million barrels.)

The main petroleum-producing regions are the Middle East, around the Persian Gulf; the United States, particularly Texas; Venezuela and Mexico; northern and western Africa; Russia; and in Western Europe, the North Sea. Saudi Arabia has the largest production capacity, being able to supply more than 10 million barrels of oil a day. Along with 12 other major oil-producing nations, Saudi Arabia belongs to the Organization of Petroleum Exporting

Left: An automatic oil pumping station in the hills of the Coast Range, north of Los Angeles in the United States. Such stations are common in this part of California.

Far right: Filling stations dispense many of the most important products obtained from petroleum. They include petrol, as fuel for cars; diesel oil, as fuel for lorries; and engine oil and grease for lubrication. This American filling station also dispenses gasohol, a mixture of gasoline (the American word for petrol) and alcohol.

Right: Inflating a hot-air balloon, using a propane gas burner. Propane is found in natural gas, from which it can readily be removed by liquefaction. It separates as a liquid when natural gas is put under pressure. Another "liquid" gas, butane, is obtained in the same way. Liquid propane and butane are sold as bottled gas.

Countries (OPEC). In the 1970s OPEC increased its prices by more than 20 times from $1.35 a barrel in 1970 to $29 in 1980.

In the oil-dependent countries of the developed nations, this price hike sparked off severe inflation, recession in industry and unemployment. It also led to renewed efforts in energy conservation, and prompted research into alternative sources of power (see page 22). Also it spurred the search for, and development of oilfields in technologically difficult areas, particularly offshore and notably in the North Sea.

Probably most new "oil strikes" will be made in even deeper offshore waters. This all adds up to more expensive production and a progressive increase in oil prices. So attention is now being turned to other massive deposits of oil that until recently have been uneconomic. They are the oil shales and tar sands (see page 15).

Natural Gas

Gas is often found in the rocks with petroleum deposits and has likewise become a major fuel. It currently supplies about one-seventh of the world's energy. It is often called natural gas to distinguish it from manufactured fuel gas, such as COAL GAS and PRODUCER GAS. The main gas in natural gas is methane. It also contains gases that are liquefied and sold as BOTTLED GAS.

The United States and Russia are the biggest natural gas producers, with a combined output of nearly 1 million million cubic metres per year. To distribute gas to Western Europe Russia has embarked upon one of the biggest construction projects in history – the 6000-km long Trans-Siberian Pipeline (see page 85). There are also vast deposits of natural gas under the North Sea, which have been tapped since the mid-1960s.

Oil and Gas 2

Above: An oil-rig worker "wrestling" with a drill bit, which has just been withdrawn from the borehole. The cutting teeth of the bit can be clearly seen.

Opposite: A diagram showing a typical North Sea oil production operation, drawn to scale.

Below: A drilling rig in the frozen wastes of Alaska. Oil prospectors are being forced into ever more remote areas in search of the precious "black gold".

Prospecting and Drilling

Oil and natural gas are found underground trapped in the rock layers. Like coal, they are fossil fuels, formed over periods of hundreds of millions of years by the decay of simple organisms – plant and animal – that lived in ancient seas. When petroleum engineers search for oil, they look for certain types of rock layers, or strata, which they know from past experience can trap oil.

They carry out surveys from the ground and from the air using a variety of instruments, and they bore into the rocks to take samples. They also set off explosions in the ground and record the waves reflected from the underground rock layers. This is called seismic surveying. When all the information is collected and analysed, a picture of the underground strata is obtained. If it indicates that oil may be present, a test well is drilled. If that strikes oil, then production wells can be drilled.

Most wells are now bored by rotary drilling. A drill bit is fixed to the end of a drill pipe and rotated so that it bores into the ground. The drill bit consists of several (usually three) toothed cutting wheels, made of hardened steel or diamond and tungsten carbide. At the top of the drill pipe is a square section called the kelly, which is gripped by the turning mechanism, the rotary table.

As drilling proceeds, more lengths of pipe are added. When the drill bit has to be changed, the whole pipe has to be withdrawn and dismantled. This is done with lifting block and tackle in the tall steel derrick, which dominates the drilling rig. With drilling now being carried out at depths down to 9 km, it is not hard to imagine what a mammoth task bit changing can be.

Mud is pumped down the drill pipes, which are hollow. This helps lubricate the bit and brings back rock cuttings. It also helps prevent the borehole collapsing, although the hole is usually lined with steel tubes as it deepens. When oil, or gas, is struck, the borehole is capped with a "Christmas tree" of valves which control the flow from the well.

Offshore Operations

It is difficult and expensive enough drilling for oil on land. Offshore, difficulty and cost rise dramatically. But many of the remaining oil deposits are to be found under the relatively shallow stretches of water that surround the continents – on the so-called continental shelves. Some offshore oilfields have been tapped for many years, in the Persian Gulf and in Venezuela. But there the water depths are quite modest, the climate is warm and the sea is relatively calm. Nowadays, however, it is becoming increasingly necessary to tap fields in much deeper water, often in appalling conditions.

Some of the worst conditions are experienced in the

rich North Sea oilfields being developed by Britain and Norway. There, oil is being extracted from water depths down to 200 metres, where the water pressure at the bottom is 20 times that of the atmosphere. Temperatures are nearly always low, winds can gust to 160 km per hour or more, and waves can climb to a height of 30 metres.

To cope with such frightful conditions new techniques and technologies have had to be developed. Drilling is often done by drilling rigs that are half-submerged, called semisubmersibles. They consist of a drilling platform mounted on huge buoyancy tanks sunk deep in the water. This design keeps the rig steady even in rough weather. It needs to be kept steady when it has to drop a drill stem through up to 200 metres of water and then bore a hole through maybe 3000 metres of rock to reach (hopefully) an oil deposit.

Into Production

If oil is found, the borehole is temporarily capped and the drilling platform moved on to search for oil elsewhere. The drilling platform is replaced by a production platform, which drills more boreholes into the oil deposit and extracts the oil. Several dozen holes are usually drilled. They start nearly vertically, but then fan out in all directions so as to cover a large area. From the production platforms the oil is taken to the mainland either through pipeline or by tanker.

There are two main types of production platform. One consists of a steel tower that is fixed to the seabed by piles. Each leg of the tower is "pinned" down by steel rods driven up to 150 metres into the seabed. The other type is made of reinforced concrete. It is known as a gravity platform because it relies on its enormous weight to stay put on the seabed. The Norwegian Statfjord concrete gravity platform weighs more than 800,000 tonnes.

Steel rigs are floated out to the oilfields horizontally and then upended into position. Concrete rigs are floated out vertically using huge tanks around their base as buoyancy tanks. When on station, the tanks are flooded, and the structure sinks.

Magnus

One of the biggest steel rigs stands in the Magnus oilfield off the Shetlands. It is a massive structure made up of 13 km of steel pipe weighing 14,000 tonnes. On top of the tower itself is the production deck. This consists of eight storeys, which include not only the drilling and oil-handling equipment but also accommodation for some 200 workers. It was assembled from 14 modules weighing up to 2000 tonnes each. The highest point of the Magnus rig stands 312 metres above the seabed, making it slightly taller than the Eiffel Tower in Paris!

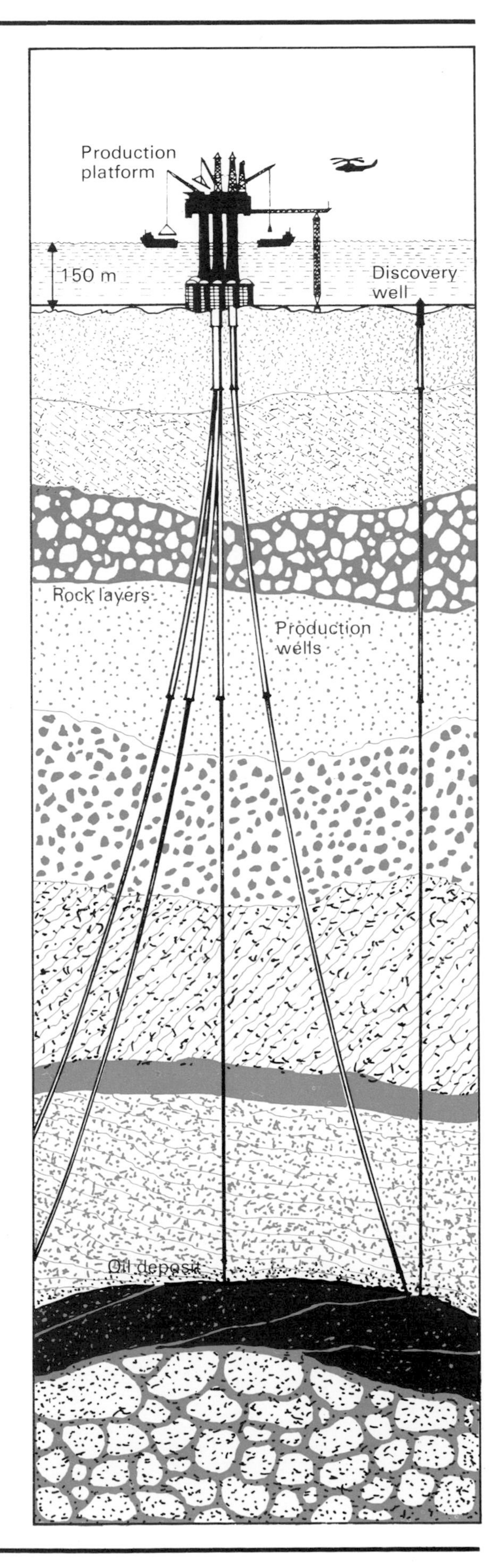

Oil and Gas 3

Saturation Divers

To support offshore exploration, new diving vessels and new diving techniques have been devised. The vessels are miniature submarines called submersibles. They work from a surface support ship, and are on hand with remote manipulator arms, lights and television cameras to assist in underwater operations such as pipe laying. The latest ones are equipped also to transport divers to their place of work on the seabed. They are called lock-out submersibles because they incorporate a pressurized air lock through which divers can leave and enter.

These divers live for up to a week under pressurized conditions. This enables them to work for long periods on the seabed without having to decompress in between, as is normally necessary in diving. They are called saturation divers because their blood becomes saturated with the oxygen/helium mixture they breathe. When they are not working, they live in a compression chamber on board the drill rig or on a support ship. They have to spend about a fortnight gradually decompressing before they can return to a normal life in the atmosphere.

Oil Mines

In oil shales and sands the oil is not found in liquid form. It is found as a sticky tar absorbed in the rocks like water in a sponge. The United States alone probably has as much oil locked in shales as can be extracted from all the ordinary oilfields in the world put together. Their main deposits are the Green River shales in Utah, Wyoming and Colorado. Canada has nearly as much oil locked in the Athabasca tar sands in Alberta.

The principles behind extracting the oil are simple: mine the shale or sand and heat it with steam. The sticky tar becomes liquid and flows out. In practice it is not so simple. However, Canadian companies have had some success with the Athabasca tar sands at Fort McMurray with two plants designed to produce up to 180,000 barrels a day. The tar sand is located about 15 metres below the surface, and is mined by ordinary opencast methods. It is then treated with superheated steam, and the liquid tar that comes off is processed to give an oil suitable for refining. It is often called syncrude (synthetic crude oil).

Many oil shales and tar sands, however, are found too deep to be mined at the surface, and different methods are needed. In some experiments steam is piped down to liquefy the deposits, and the hot liquid oil is then piped back to the surface. In others the deposits are broken up and heated underground, and the vapours given off are piped back to the surface. In the coming years oil engineers will have to rely more and more on such novel techniques to put off the inevitable day when the oil tanks of the world run dry.

Opposite: Oil-production rigs in the North Sea as night falls. The beauty of the scene belies the hazards and hardships encountered in winning off-shore oil. The two rigs are located north-west of Aberdeen in the rich Forties field, which can produce up to half a million barrels of oil a day.

Below: Laying a pipeline from a barge in the North Sea oilfields. The pipeline is constructed from welded steel pipes up to 80 cm in diameter. After welding, the pipeline is painted with tar and then given a coating of reinforced concrete. This helps weigh it down and protects it from accidental damage.

Power Stations

Electricity is the form of power most used in modern society. It is very convenient, instantly available at the flick of a switch. It gives us efficient artificial light at night-time. It drives the electric motors that keep the wheels of industry turning and that power the many labour-saving gadgets and appliances we use in the home. It powers television sets and milling machines, computers and locomotives, giant lathes and vacuum cleaners. We would indeed be lost without electricity.

For most consumers in the developed countries, electricity is available from the "mains", through cables wired into the home and factory. The source of the electricity is sometimes hundreds of kilometres away, at a power station. In two out of every three power stations in the world, the electricity is produced by means of turbogenerators driven by steam. In the others, hydroelectric power schemes, the electricity is produced by turbogenerators driven by flowing water (see page 20).

Steam Power

A turbogenerator is a combination of a turbine and an electricity generator. In a steam turbogenerator high-pressure steam is used to spin the turbine. This turns the coupled generator, and electricity is produced. In power stations the steam is "raised" by heating water in a boiler. The heat is usually supplied by burning fossil fuel – coal, oil or natural gas. In a nuclear power station, however, the heat comes from a reactor, which uses uranium as "fuel" (see page 18).

Steam turbines are among the world's most powerful machines. Coupled with their generator on a single shaft, they can stretch for more than 50 metres. A machine as big as that can produce nearly 700,000 kilowatts, enough to supply a large town.

A steam turbine consists basically of a shaft ("rotor") turning in a fixed casing ("stator"). The rotor carries a

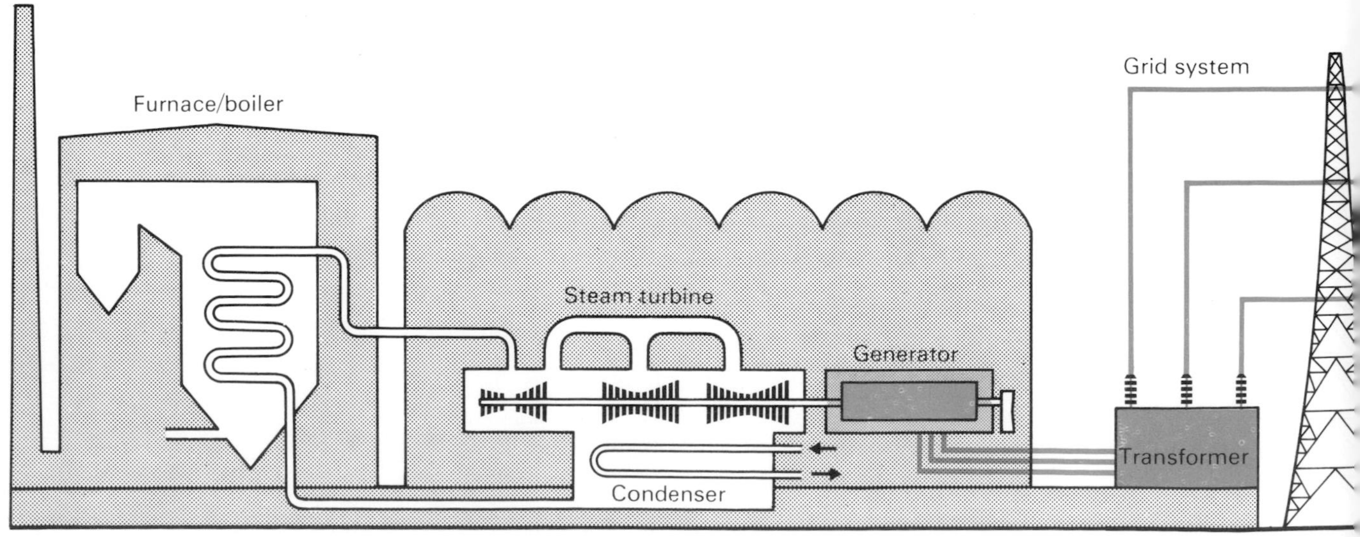

Above: The central control room at a nuclear power station. The complex array of dials, instruments, gauges and recorders continuously monitor the operations in all parts of the station.

Left: A giant steam turbine being installed in a power station. The rotor consists of five separate turbine units joined together. Steam is introduced first to the small, high-pressure turbine (in the front). It then goes to the larger, lower-pressure turbines in turn.

Below: An outline of the production and distribution of electricity, from power station to consumer.

number of discs made up of many separate blades ("vanes"). When steam passes through the vanes, the rotor spins. For efficiency, several sets of vanes, or stages, are used of gradually increasing diameter. High-pressure steam is piped through each stage in turn before passing into a condenser.

In the condenser the steam is cooled by pipes containing cold water and condenses, or changes back into water. Thus creating a partial vacuum, which draws more steam through the turbine. The condensed water then goes back to the boiler to be turned into steam again.

Generation and Transmission

The usual type of electricity generator used in power stations is called an alternator. It produces two-way, or alternating current (AC), at a voltage of about 25,000 volts. The electricity then has to be sent, or transmitted to the consumers – the people who use it. The most efficient way of doing this is to send it through cables (transmission lines) at very high voltage, often at 400,000 volts or more. Fortunately, the voltage can be increased quite easily, by means of a TRANSFORMER.

The transmission lines consist of sets of three cables. Usually they are carried cross-country on tall steel towers, or pylons, and insulated from the towers by glass insulators. The lines are usually made of aluminium strengthened with a steel core. They carry the electricity to substations which distribute it to the consumer.

To even out the supply of electricity over a region, or even a country, a number of power stations are linked in a so-called grid system. Then power can flow wherever it is needed most at any one time. Britain has the largest grid system in the world, which is also connected by undersea cable to the grid system of France. The present capacity of 160,000 kilowatts will expand to over 2,000,000 kilowatts in 1986.

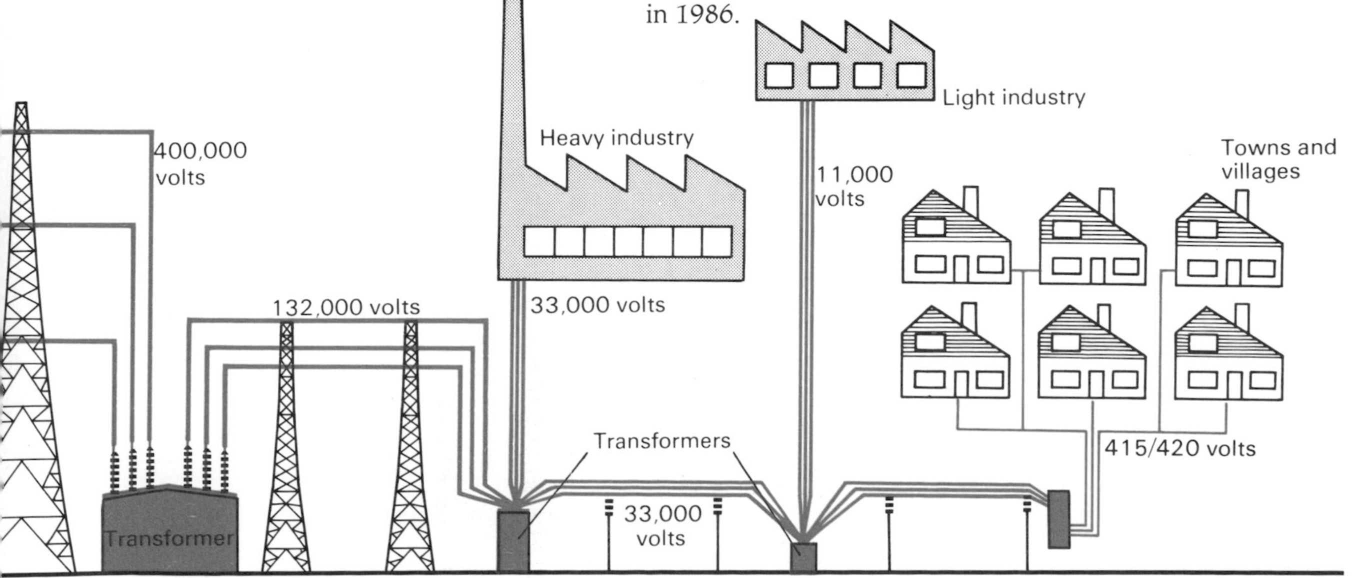

Nuclear Power

In 1956 a new power station at Calder Hall, in north-east England, began feeding electricity into the national grid. It was different from all earlier power stations: it used energy released from inside atoms, or rather the nucleus (centre) of atoms. It was the first nuclear power station. Today there are more than 200 nuclear power stations throughout the world. The United States, Russia, Britain and France have the most.

In a nuclear power station, heat is produced when atoms of the heavy metal uranium are made to split. This process is called NUCLEAR FISSION. The energy released during fission is enormous. A few kilograms of uranium undergoing fission release as much energy as the explosion of tens of thousands of tonnes of the explosive TNT.

This enormous energy is released in a fraction of a second in the awesome ATOMIC BOMB. In a nuclear power station, however, it is released very slowly and under careful control. Fission takes place in a nuclear reactor. Heat from the reactor is used to heat water into steam, then the steam is fed to turbogenerators to produce electricity in the ordinary way (see page 16).

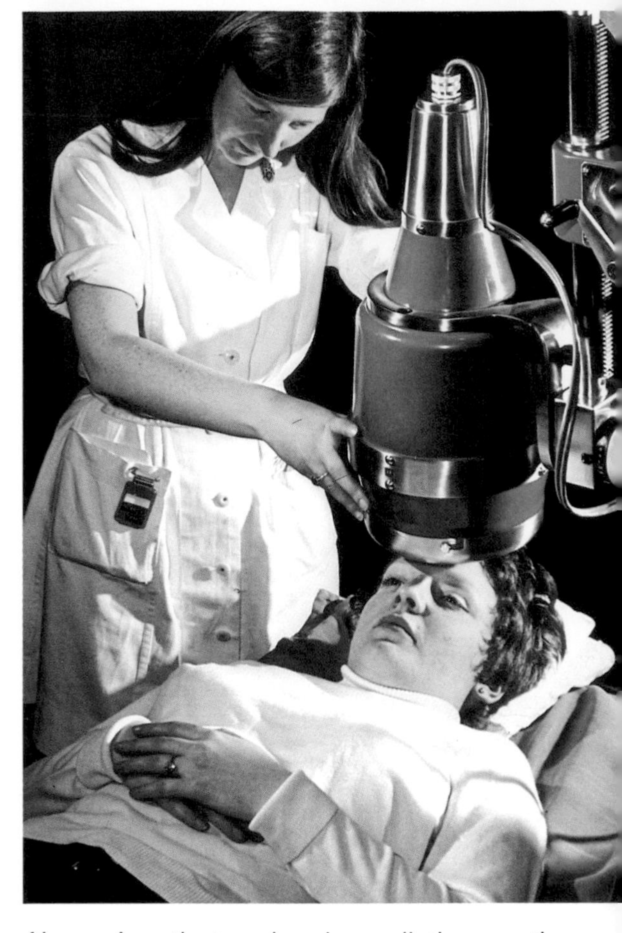

Above: A patient undergoing radiotherapy, the treatment of disease by radiation. Most sources of radiation used are produced in nuclear reactors. They are known as radioisotopes.

Nuclear Reactors

A nuclear reactor is a large, often spherical vessel, constructed from reinforced concrete and steel. The uranium "fuel" that is to undergo fission is contained in fuel rods in the reactor core. A coolant (cooling fluid) circulates through the core and takes away the heat produced by fission. It then goes into a steam generator, where it heats up water into steam. The steam is then led off to the electricity generators. The coolant returns to the reactor core to extract more heat.

In the most common kind of reactor, used in more than 20 countries, the coolant is water under pressure. This is known as a pressurized-water reactor (PWR). In other reactors carbon dioxide gas is used as a coolant. These reactors also contain graphite in the core.

The fission reaction is controlled by means of control rods, which are pushed in or taken out to make the reaction go slower or faster. They are designed to "fail-safe". In any emergency they are pushed right in and shut down the reactor.

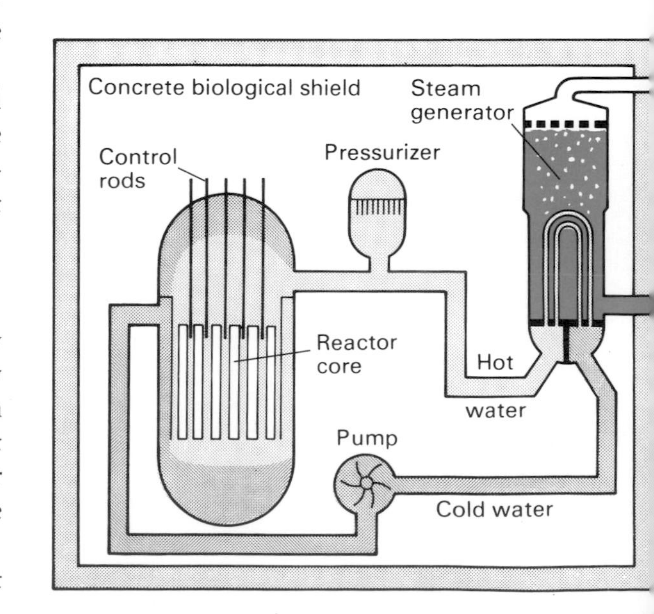

Breeder Reactors

A new type of nuclear reactor called the fast reactor is now coming into use. It is much more efficient than an ordinary reactor because it can extract twice as much energy from its fuel. France has built the first commercial fast reactor at Creys-Malville in south-west France. It is called the Super Phenix. Britain and Russia have also had considerable experience in operating fast reactors.

A fast reactor uses liquid sodium metal as a coolant. It

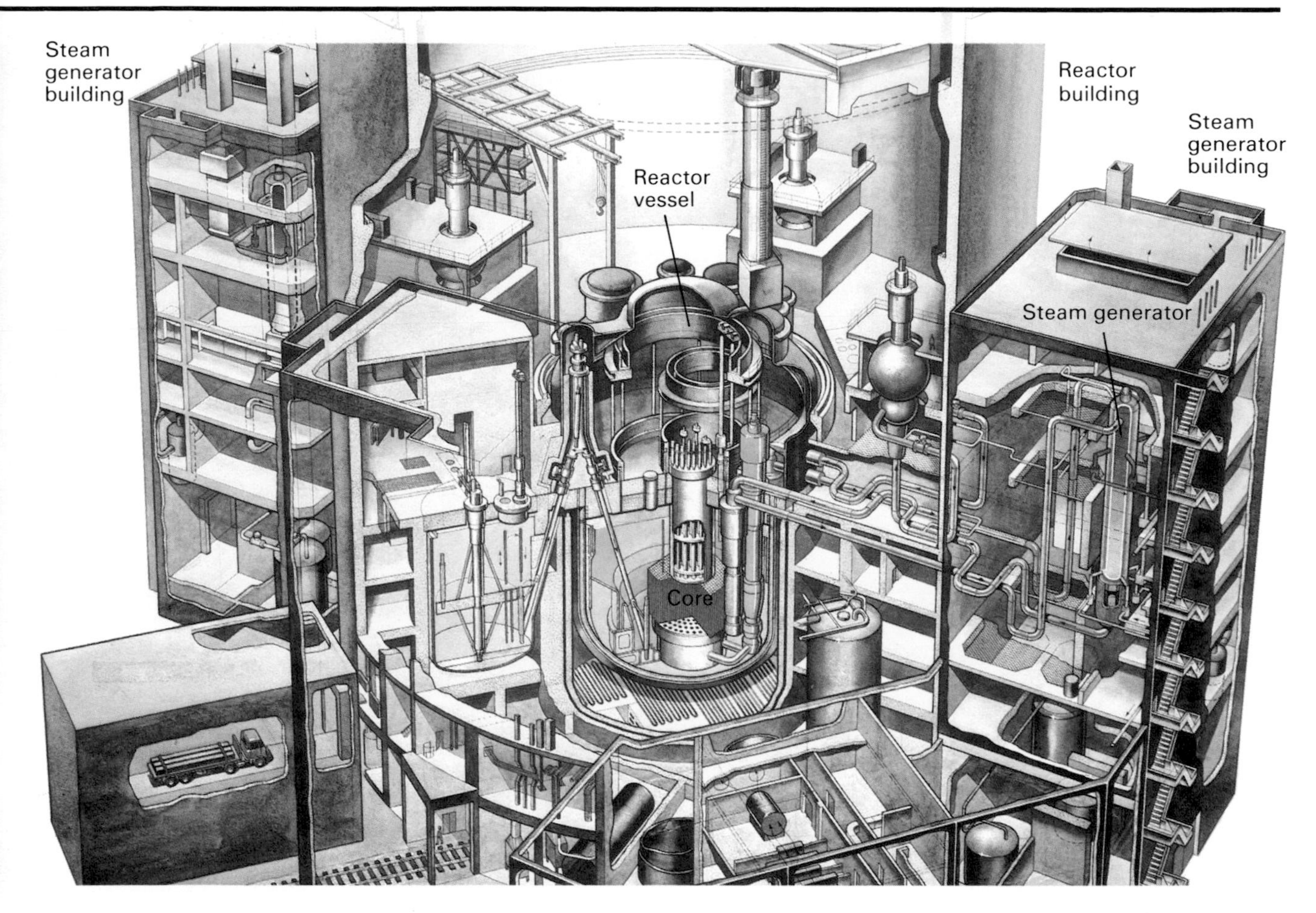

Above: A detailed cutaway of the Super Phenix, France's fast reactor at Creys-Malville. It has a maximum power output of 1200 megawatts (million watts). It uses liquid sodium as a coolant.

Below: The essential features of a nuclear power plant using a pressurized-water reactor. The coolant extracts heat from the core and circulates through the steam generator. There, water turns into steam to drive the turbogenerator.

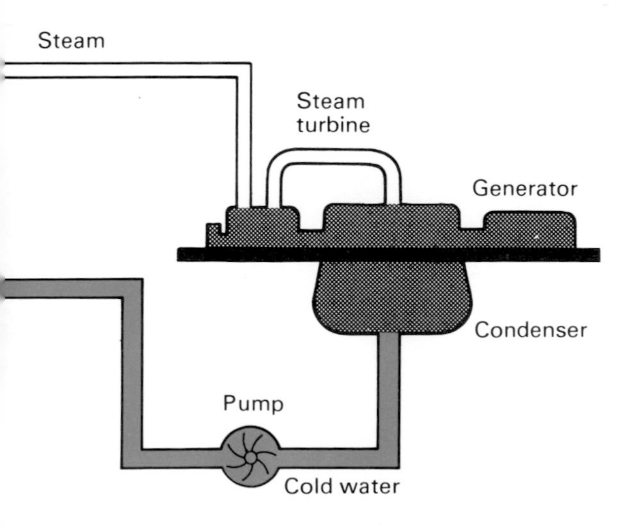

is also known as a breeder reactor because it breeds new fuel. The reactor core is surrounded by another layer of uranium. Radiation produced by fission converts some of this uranium into plutonium, which is another excellent nuclear "fuel".

Danger: Radiation!

Nuclear power plants suffer from one major problem, however. Nuclear fuels such as uranium and plutonium are RADIOACTIVE: they give out dangerous and very penetrating radiation. Even more radiation is produced during fission. Anyone exposed for only a short while to this radiation becomes sick and may die. The radiation attacks living tissues. It is harmful even in quite small doses because it can alter the genes in body cells. Such MUTATION can affect later generations.

For this reason nuclear reactors are built with thick walls that prevent radiation escaping. All nuclear materials are shifted in lead containers and handled by robot arms. Another great problem is posed by the radioactive wastes that reactors produce. They remain dangerously radioactive for hundreds or perhaps thousands of years. At present wastes are often stored in underground tanks, or sealed in containers and dropped into deep ocean trenches. Neither method is very satisfactory.

Hydroelectric Power

The Romans were great engineers, building fine roads, bridges and aqueducts. They also developed the waterwheel to harness the power of flowing water. The waterwheel remained the main mechanical power source for some 1700 years until the development of a reliable steam engine. But water power is still being harnessed today by means of the modern version of the waterwheel – the water turbine. It is now used to produce electricity known as hydroelectricity ("water-electricity").

The biggest power stations in the world are hydroelectric. Russia has built several large hydroelectric power (HEP) stations in Siberia, where fast-flowing rivers abound. The one at Krasnoyarsk on the River Yenisey can generate over 6000 million watts (or megawatts, MW) of electric power. This is twice as much as the biggest steam power station. There is a massive HEP plant at Itaipu on the River Parana on the Paraguay/Brazil border in South America. When fully operational, it will supply up to 12,000 MW.

On average, HEP stations provide about one-fifth of the world's electricity. But in very mountainous countries, including Norway and Switzerland, they supply nearly all the electricity. HEP stations are usually expensive to build, but they have the great advantage that their "fuel" – flowing water – is free.

Producing Hydroelectricity

To generate hydroelectricity you need a "head" of water, or a difference in level. Water can then flow from the high to the low level. A power station is built at the lower level and water flows at high speed through water turbines and spins them round. The turbines are coupled to generators, which produce electricity when they spin.

Most water turbines look much like ships' propellers. They have several blades, or vanes, set at an angle. On many the angle can be altered, depending on the quantity of water flowing through. Some water turbines are more like the ancient waterwheel. They consist of a wheel with buckets around the edge. The water is directed at the buckets by a nozzle and spins the wheel round.

A mountain lake may sometimes provide a natural "head" of water for a HEP scheme. The water is then piped to a power plant lower down the mountainside. But often, an artificial lake, or reservoir, is created by damming a valley or a river gorge (see page 80). The HEP plant may then be located at the foot of the dam or even inside it.

Pumped Storage

When a hydroelectric plant is operating, the level of water in the high-level lake or reservoir naturally falls. In an ordinary hydroelectric scheme, the level can only be restored by natural means, such as rainfall. In summer, the level may become very low indeed.

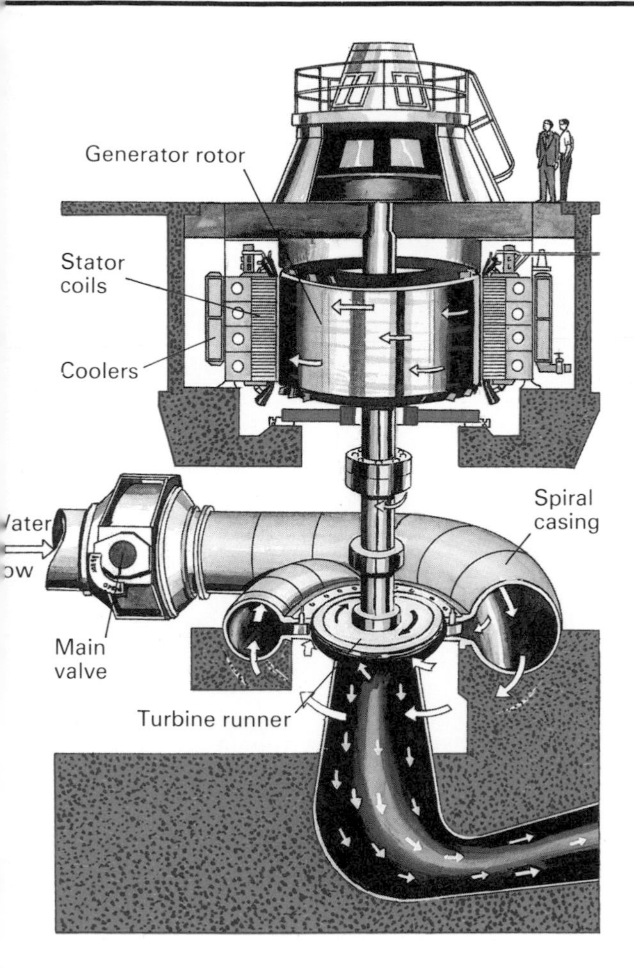

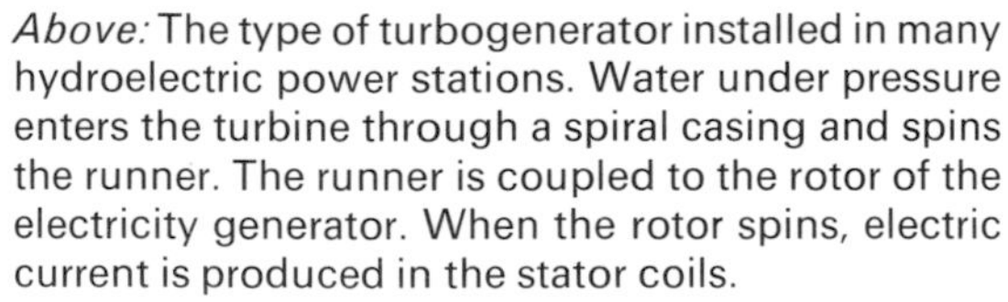

Above: The type of turbogenerator installed in many hydroelectric power stations. Water under pressure enters the turbine through a spiral casing and spins the runner. The runner is coupled to the rotor of the electricity generator. When the rotor spins, electric current is produced in the stator coils.

Top right: A vast cavern being excavated to house the turbogenerating machinery for the Dinorwic pumped-storage power station in Wales. The power station uses six turbogenerators and has a maximum power output of 1880 megawatts. This is more than the average power consumption of the whole of Wales.

Left: Part of the mammoth hydroelectric and irrigation scheme completed in the 1970s in the Snowy Mountains, in south-east Australia. The scheme involved diverting the waters of the Snowy River, which naturally flows south, back through the mountains towards the arid western plains. The Snowy Mountains scheme required the building of 17 large dams and many small ones, nine power stations and some 300 km of tunnels and aqueducts.

For this reason some hydroelectric schemes are designed for pumped storage. The biggest in Europe is at Dinorwic in North Wales. A pumped-storage scheme uses two reservoirs – one at high level, the other at low level. During the day when power demand is high, water is allowed to flow from the high to the low reservoir through the turbines, generating electricity. During the night, when power demand is low, the turbines are turned into pumps and pump water back from the low to the high reservoir. The original head is restored, and maximum power is again available next day.

Tidal Power

In some parts of the world the difference in water level between high and low tides is also used as a "head" to generate hydroelectricity. The most successful tidal power plant, on the River Rance in Brittany in north-west France, opened in 1966. In the Rance estuary there is a difference of over 10 metres between low and high tides. A low-level dam, or barrage, across the river holds back the water until there is enough "head" to work the 24 turbines.

Proposals for much larger tidal power schemes have been put forward for the Bay of Fundy in Canada and the Severn Estuary in Britain. They would be very expensive to build but have very low operating costs.

Power for Tomorrow 1

One of the biggest problems facing the world is that of energy shortage. Most of the energy the world uses today comes from fossil fuels: coal, oil and natural gas. These fuels have formed over hundreds of millions of years, and once they have been used up they cannot be replaced. Although there is enough coal to last for several centuries, supplies of oil could run out within fifty years.

Other important sources of energy at present include nuclear and hydroelectric power. Nuclear power depends on a supply of uranium, which comes from minerals. These minerals could become scarce early next century, so nuclear power cannot provide the world with energy for long.

For several countries the further development of hydroelectric power will help overcome the energy shortage. But in most countries the land is not suitable for large-scale hydroelectric power production. But hydroelectricity does point the way to the future, because it is a renewable source of energy. As long as rains fall, rivers flow, and tides ebb and flow, there can be hydroelectricity.

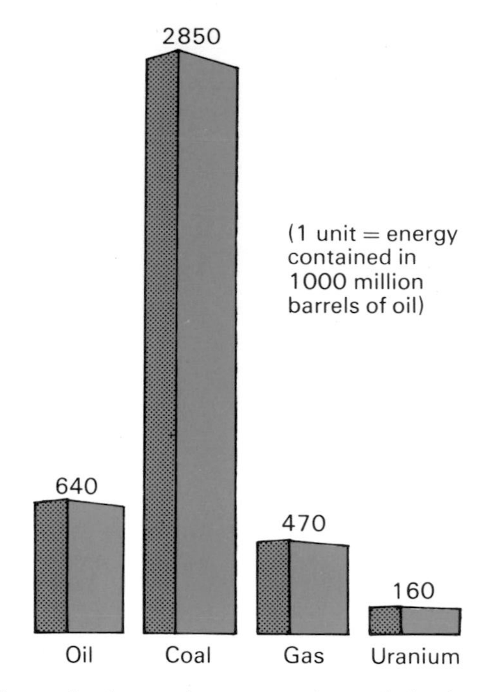

Above: An interesting comparison of the known reserves of the major fuels in terms of the energy they would provide. As can be seen, coal far outstrips the rest. Used at its present rate, coal should last for at least another 200 years.

Renewable Sources

As time goes by it will become increasingly necessary to switch to renewable sources of energy. The most abundant source is the energy from the Sun, or solar energy. The Sun bathes the Earth in 20,000 times more energy than we currently use. Much of this energy goes into the Earth's weather system – into heating the land and oceans and driving the wind and the waves. In sunshine, wind and ocean waves we have unending renewable sources of energy.

Geothermal Power

There is another renewable source of energy in the rocks that form the Earth's crust. The deeper we bore into the ground, the higher the temperature. In certain volcanic regions, hot springs and geysers make this rock heat, or geothermal energy, readily available. It is already being harnessed. Certain districts of Paris and Reykjavik (Iceland) are heated by hot springs. In Italy, Japan, New Zealand and the western United States underground steam is used to generate power. The geoethermal power plant at Wairakei, in New Zealand, generates about 7 per cent of the country's electricity.

Experiments are also underway to harness "hot rock" energy in non-volcanic areas. In one scheme, tried successfully at Los Alamos in the United States, two holes several kilometres deep are drilled into hot rock. Water is pumped down one hole under great pressure, fracturing the rock. It passes through the shattered rock and, now piping hot, flows up the second borehole to the surface, to be used for heating.

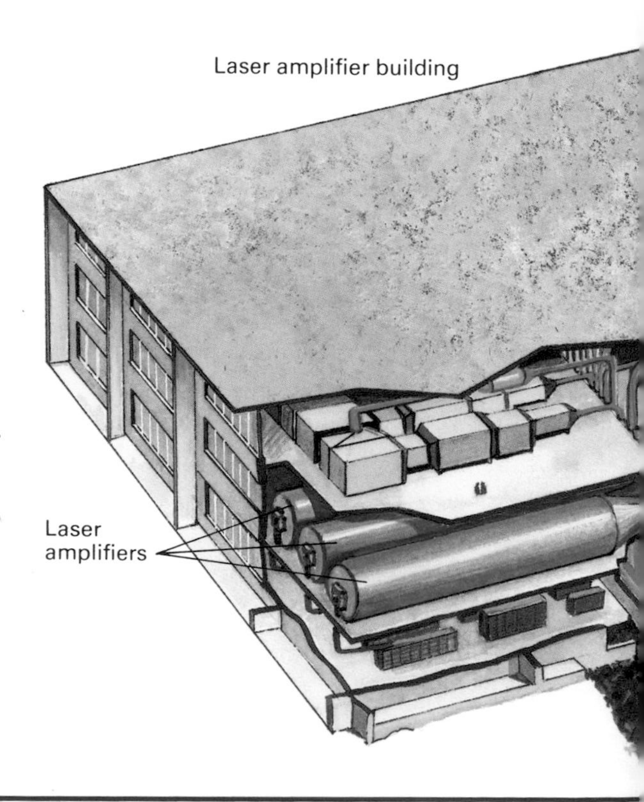

Right: A geothermal power plant in the hills of northern California, a region of natural geysers and hot springs. Steam is tapped underground and led through pipes to steam turbogenerators. The steam visible in the photograph, however, is coming from the power plant's cooling towers.

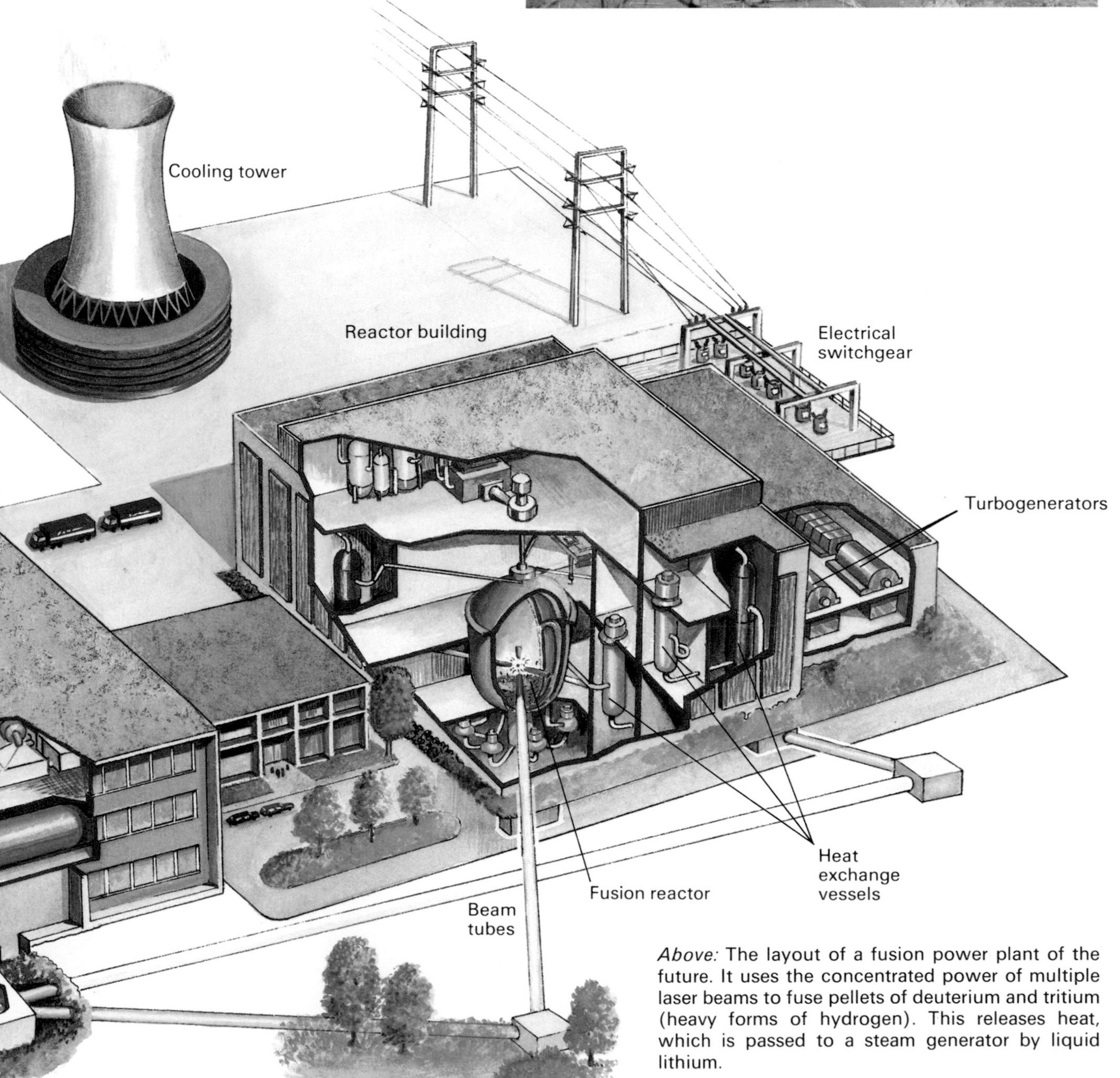

Above: The layout of a fusion power plant of the future. It uses the concentrated power of multiple laser beams to fuse pellets of deuterium and tritium (heavy forms of hydrogen). This releases heat, which is passed to a steam generator by liquid lithium.

Power for Tomorrow 2

Above: A solar-power satellite in high Earth orbit. This particular design uses long arrays of solar cells to capture the Sun's energy.

Below: Solar panels cover the roof of the George A. Towns School in Atlanta, Georgia. They provide all the energy for space heating in winter and cooling in summer.

Solar Power

In many countries solar power is already being tapped for domestic hot-water heating. This is done by means of flat-plate collectors; panels that trap the Sun's heat and use it to heat up water circulating through pipes. They naturally work best in sunny climates, but function quite well even in cloudy conditions.

In several countries scientists are now experimenting with solar power schemes to produce electricity on a commercial scale. Near Albuquerque in New Mexico, American scientists have built an experimental solar "power-tower" to concentrate and collect the Sun's rays. It consists of hundreds of mirrors ("heliostats"), clustered around a tower 60 metres high. They reflect sunlight onto a boiler at the top of the tower, which heats water to steam. In a working power-tower station, as many as 20,000 mirrors would be used, and the steam produced would be fed to turbogenerators to produce electricity.

Solar power stations on the ground would not produce electricity at night, of course. So some scientists have suggested building them up in space, where the Sun shines all the time. Such satellite solar power stations could take the form of vast reflecting dishes several kilometres across. The energy they produce would be beamed down (as MICROWAVES) to collectors on Earth and converted there to electricity.

Alternatively, the satellite power station might use long arrays of solar cells to produce electricity. Solar cells already power most spacecraft, but at present they are too inefficient and too expensive to be used on a large scale.

Above: A new type of wind-power machine being tested at Sandia Laboratories in Albuquerque, New Mexico. Known as a vertical-axis wind turbine, it is a particularly efficient design, which can accept wind blowing from any direction.

Below: A cutaway of the tokamak designed by European nuclear physicists, known as the Joint European Torus (JET). It should be capable of reaching temperatures of tens of millions of degrees.

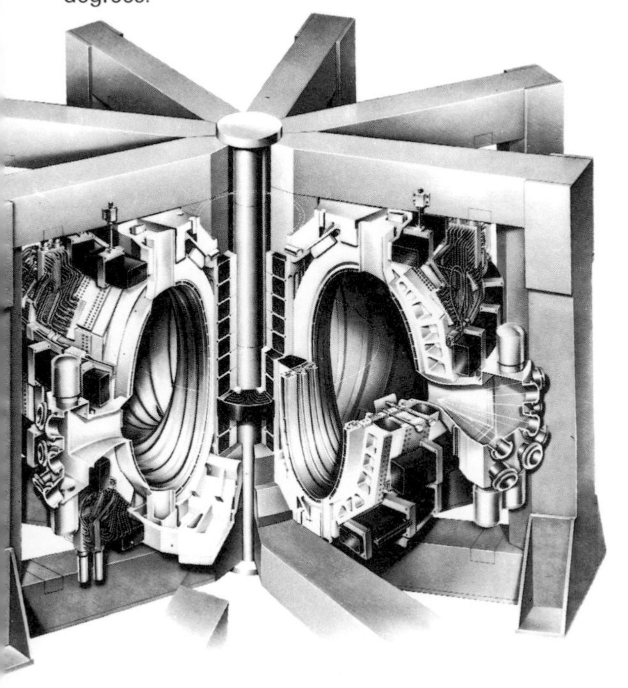

Wind and Wave

The energy blowing in the wind has been harnessed for centuries, using windmills, to grind corn, pump water and more recently to generate electricity. Hundreds of isolated farms still rely on the modern version of the windmill – the wind turbine – for their electricity.

Much larger wind turbines are now being built to harness the wind for large-scale production of electricity. Some experts foresee the building next century, perhaps offshore, of wind farms comprising many huge turbines.

One of the largest present-day machines is at Boone, in North Carolina, in the United States. It consists of a twin-bladed propeller, 60 metres long, on top of a steel tower. Another large machine has been built at Tvind, on the west coast of Denmark. Both can produce about 2 million watts of electric power.

Several devices have also been invented to tap the energy in the waves. They include duck-like floats and rafts that produce power when they nod up and down with the waves. But they would have to extend over several kilometres of sea in order to produce a reasonable amount of electricity.

Fusion Power

These renewable sources, however, will not be able to supply all the energy required for mankind to continue his technological progress. We must look to another source of energy – NUCLEAR FUSION. This is the process which keeps the Sun and the stars shining. Scientists have already harnessed fusion energy for destruction, by developing the terrible HYDROGEN BOMB. Now they are seeking methods of controlling fusion for peaceful purposes.

Nuclear fusion takes place when atoms of heavy hydrogen combine to form atoms of helium. When this happens, large amounts of energy are given out. The problem is that this fusion occurs only at temperatures of many millions of degrees. So somehow the hydrogen must be heated to such temperatures and held in some kind of "container" that can exist in this heat.

Scientists have in fact been able to do both, by means of powerful electromagnets. The machine they use is called a tokamak. European fusion scientists are working on a powerful tokamak called JET ("Joint European Torus") at Culham, in Berkshire, England. The Americans, Russians and Japanese also have large tokamaks which could one day lead to a practical fusion reactor and power station. Other scientists are experimenting to try to bring about fusion by means of laser beams.

If fusion energy could be harnessed, it could banish for ever the threat of energy shortage. The hydrogen "fuel" for fusion reactors is readily obtained from the vast oceans that cover more than two-thirds of our planet.

MASSEY FERGUSON
MF
760
MF

Chapter 2

Food and Farming

Until about 200 years ago most people lived in small communities and farmed the land. These communities were largely self-supporting. But during the Industrial Revolution people left the countryside in huge numbers to work in factories in towns. To make up for the lack of labour, the farmer turned increasingly to machines.

Today, farming in many countries has become a highly mechanized industry in which fewer and fewer people are producing more and more food. Science has also come to the aid of the farmer in providing higher yielding varieties of crops and chemical fertilizers and pesticides.

Advances in food processing and preservation have made a wider range of foods available to everyone. Artificial foodstuffs are also finding their way into the family diet.

Nothing sums up the achievements of modern farming better than the combine harvester. It is a mobile grain-producing factory constructed of some 30,000 parts. About three million combines are now in use worldwide.

Farming the Land

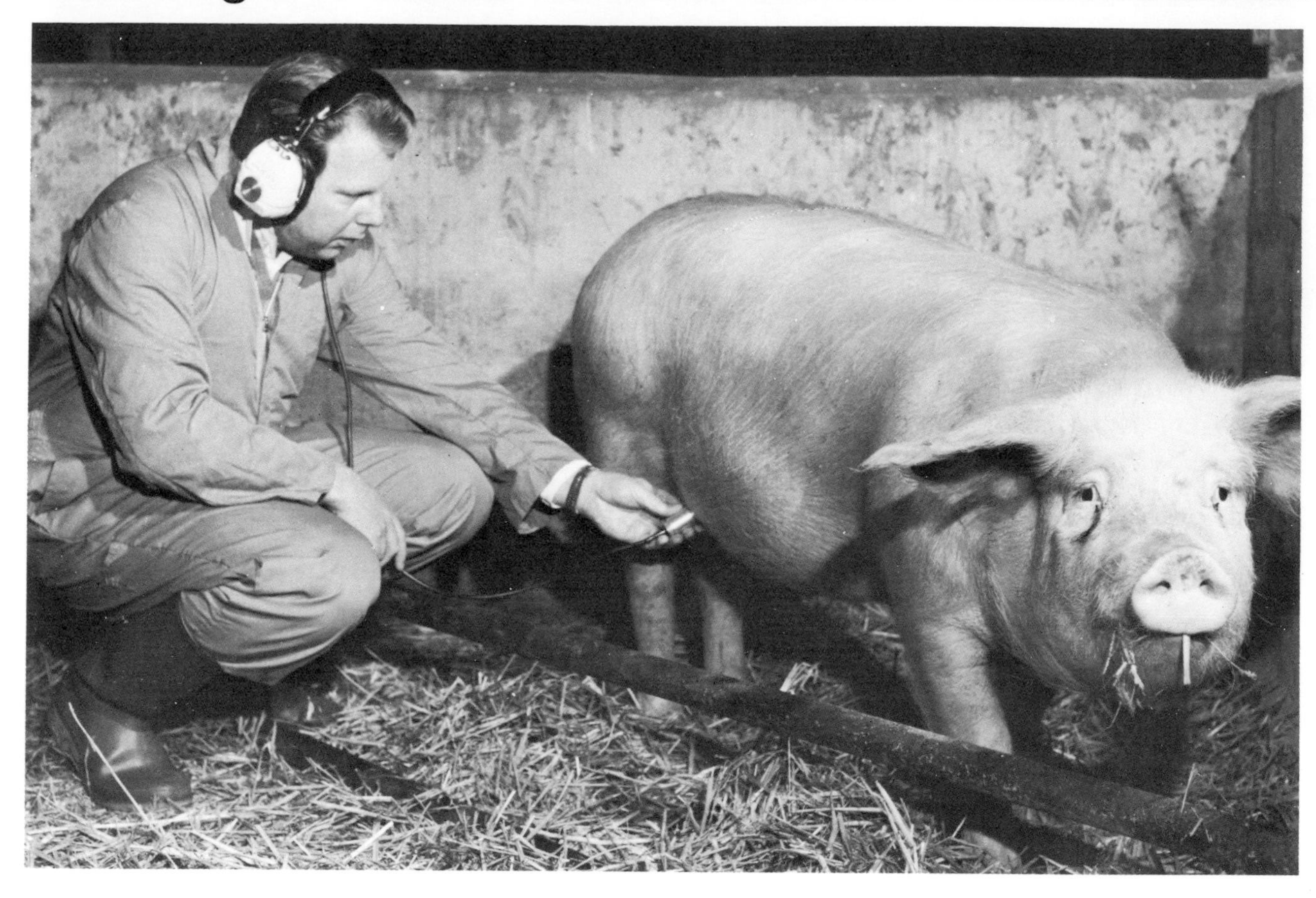

Above: A farmer uses an ultrasonic scanner to test whether a sow is pregnant.

Until about 10,000 years ago people were hunter-gatherers. For food they hunted wild animals and gathered fruits, nuts and roots. Then they discovered how to domesticate animals and to plant crops. They became farmers. This meant that they could settle in one place instead of roaming far and wide in search of food.

Farming also produced a food surplus. One person could produce enough food for several. This led to the development of a more advanced society in which people had time to expand their interests and develop specialist skills. In other words the coming of farming was the key to the development of civilization.

One of the most common methods of farming practised over the centuries has been "slash and burn". Land is cleared by slashing down and burning the vegetation, and crops are grown. After a few years, however, the soil becomes poor, and the farmer moves on to slash and burn elsewhere. This form of shifting cultivation is still followed in some countries even today. However, in developed countries farming has become an efficient industry. Like all modern industries it relies greatly on machines, particularly the tractor and combine harvester (see page 32).

The Green Revolution

Since 1950 crop yields have improved in leaps and bounds. The increase has been most dramatic in the

Below: By spraying his fields with chemicals the farmer greatly improves crop yields.

Above: Forestry must be treated as a farming, rather than as a mining operation. After mature trees have been felled, they must be replaced with seedlings to ensure continuation of timber supplies.

production of cereal crops such as wheat. Wheat yields over this period have virtually doubled. The United States alone produces nearly 65 million tonnes a year. The massive increase that has occurred in crop production has been termed the Green Revolution.

The Green Revolution has resulted from the introduction of improved crop varieties and the liberal use of fertilizers and pesticides. Crop varieties, particularly of cereals, have been improved by careful selective breeding and by artificial MUTATIONS. Scientists bring about these mutations by irradiating plants with gamma-rays. The rays alter the genetic make-up of the plant – the way they reproduce. This leads sometimes to varieties better in some way than the original.

The use of fertilizers increases crop production by continuously replacing the nutrients that plants take out of the soil when they grow. Fertilizer manufacture is now one of the biggest branches of the chemical industry. A form of agriculture called HYDROPONICS expands the idea of fertilizers by growing plants without soil in a chemical solution. Pesticides are other chemical products also produced in vast quantities to control insects, combat disease and keep down weeds (see page 60).

However, the over-use of any chemicals can have a disastrous effect on the environment. The over-use of fertilizer can lead to a form of pollution known as EUTROPHICATION, which kills rivers. The over-use of pesticides is even more harmful and poses a hazard to all forms of life (see page 68).

Factory Farming

Increases in productivity have also occurred in the other main branch of farming – livestock farming. Some livestock, particularly pigs and poultry, are now reared by what are called factory farming methods. Throughout their lives they are kept indoors under controlled conditions and are fed a carefully balanced diet. Feeding and waste disposal is done automatically. In some pig-rearing units in eastern Europe 10,000 animals or more may be reared together in this way. But the biggest application of factory farming is in the raising of chickens for meat and, in "batteries", for egg production.

A great all-round improvement has also taken place in livestock breeds. In cattle particularly it has been made possible by the technique of ARTIFICIAL INSEMINATION. Success has also recently been achieved by what is known as embryo transplanting. Particularly fine cows, for example, are treated with hormones and produce several embryos. These are then transplanted to a number of ordinary "foster" cows, which eventually give birth to fine offspring. Eventually, it is thought that breeds will be improved by GENETIC ENGINEERING.

Fishing

Above: A trawler heads for the fishing grounds at sunset. Like most modern trawlers it has its net-handling gantry at the stern.

Opposite: Thousands of fish cascade onto the deck of a factory ship as the net is emptied. They disappear down a chute that leads to the fish preparation room, where they will be cleaned and put into cold storage.

Below: One of the latest fish-finders. It is a type of echo-sounder that detects shoals of fish by means of sound waves. The sound "echoes" reflected from the shoals are displayed on the colour video screen.

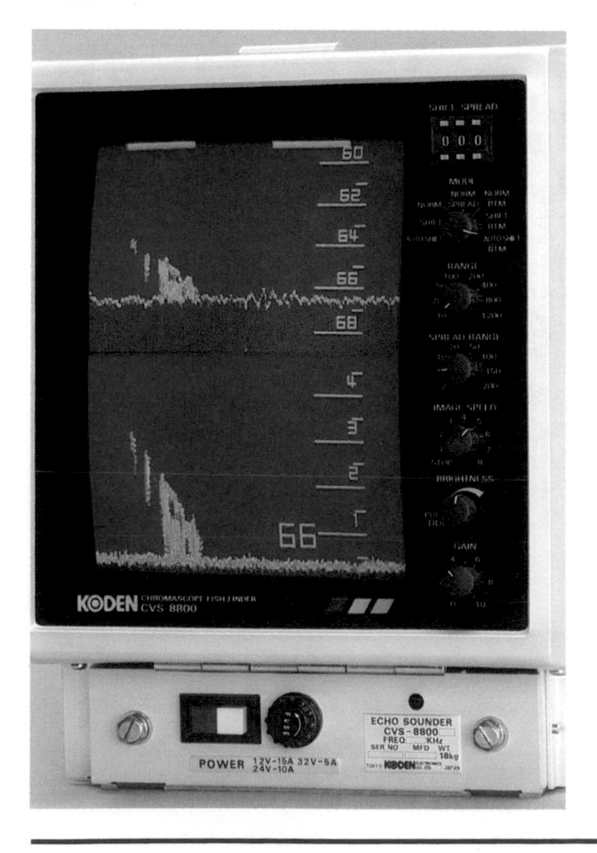

Meat from livestock provides much of the essential body-building proteins for the world's population. So does fish. Like the farmer, the fisherman has benefited from modern technology in many areas, even in the nets he uses. Many nets now are made from synthetic thread, or monofilament ("monofil"). They are much tougher and stronger than ordinary nets; have greater elasticity, or give; and never rot. When suitably dyed, they are virtually invisible in the water, which means the fish cannot see them – until it is too late. Monofil nets are estimated to have eight times the catching efficiency of ordinary nets.

Most fishing boats are now fitted with electronic fish-finders, which are types of ECHO-SOUNDERS and work by sonar. The latest ones show the presence of shoals of fish on a colour video display unit. The shade of colour indicates how many fish are present.

One of the most widely used methods of mass-catching fish is trawling – towing an open-ended conical net, or trawl, through the water. Many trawlers are now equipped as fish-factories. As soon as the catch is hauled on board it is gutted, filleted and put into deep-freeze. The smaller fish and waste are made into fish oil or fish meal for livestock feed. Even larger than the factory trawler is the factory ship, which handles the catch from many boats.

The appearance of the factory ships has underlined the problem with modern fishing methods. They are too efficient and are taking fish from the seas at too rapid a rate. Many species are dwindling in numbers in the traditional fishing grounds. Conservationists are warning that even the present annual catch, of about 65 million tonnes of fish, is too high for stocks to replenish themselves naturally.

Aquaculture

One way around the overfishing problem is to start catching species of fish that are not at present widely eaten or processed. Trawlers are already catching krill in Antarctic waters. These are the tiny shrimp-like creatures on which whales feed. They can be found in enormous swarms, and represent a very rich source of protein, if not for human consumption then for animals, or for other fish on fish-farms.

Fish-farming, or aquaculture, has been practised in the Far East for thousands of years, and China and Japan are two of the leading aquaculture nations. Japan produces nearly 100,000 tonnes of fish annually by this means. In a typical aquaculture operation the spawning, fertilization and hatching of fish eggs are carefully controlled in fish "nurseries". The young are then looked after until they are big enough to survive well on their own. Subsequent rearing may take place in artificial ponds or natural lakes, and with or without additional feeding.

Farm Machines 1

Until about 200 years ago farming everywhere required a huge labour force. In the United States at one time, two-thirds of the population were farmers. And in some countries even today, the majority of people work on the land. But in the advanced industrial societies of Western Europe and North America only a handful of people now are farmers – perhaps as few as three persons in every hundred. Yet each of them produces enough food for as many as 80 people.

The great increase in food production during the past two centuries has been brought about largely by mechanization – the use of machines. Other factors, such as the use of fertilizers and pesticides, have also played their part.

Above: A round baler being used on a Belgian farm. Moving belts inside the machine form the layer of hay or straw picked up from the ground into a roll, which is then ejected.

Machine Milking

On the dairy farm, milking machines have been used for many years. From the beginning of this century they have been of the suction/pulsation type. Cups are placed round a cow's teats and apply a rhythmic squeezing action to them by means of air pressure.

As the size of a dairy herd grows, the twice-daily milking, even with machines, becomes a very time-consuming operation. This has prompted in recent years the widespread introduction of the rotary milking parlour. The cows are milked in stalls arranged in a circle on a rotating turntable. They enter and leave the stalls one by one as the turntable rotates.

Some parlours, rotary or otherwise, have become automated. Each cow wears a collar with an electronic device attached. This sends out signals to a computer when the cow enters the parlour. The computer identifies the cow and instructs a food dispenser to deliver the correct rations to it, and later records its milk yield. All of which is added to the mounting data about the cow in the computer's memory, which the stockman can retrieve and display on a screen at the touch of a button.

Below: The cereal farmer's best friend, the combine harvester. The cut crop is threshed in the concave, and the grain is separated from the chaff by sieves.

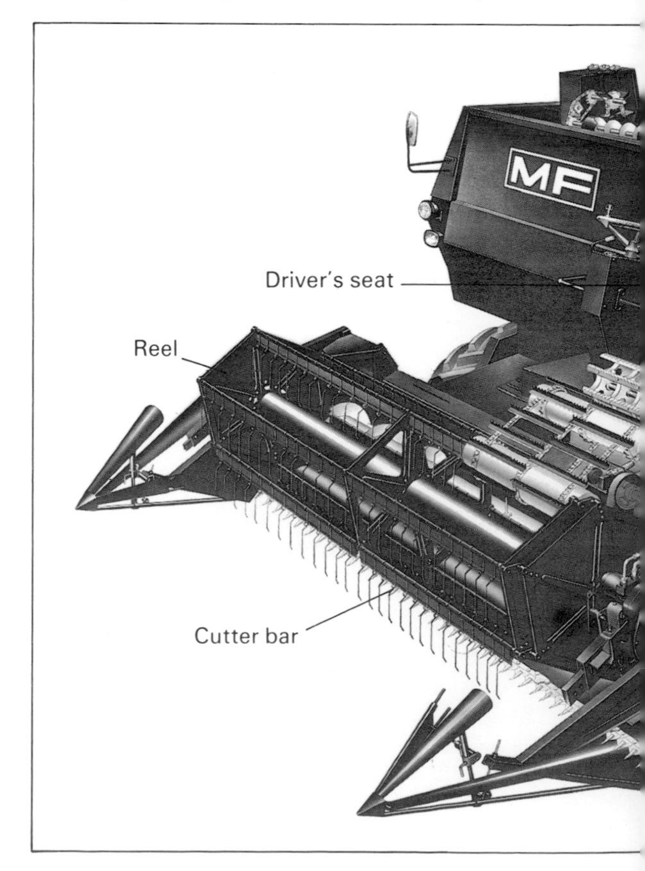

The Tractor

The most essential machine on any modern farm is the tractor, which took over from the horse as the "farmer's best friend". This is a tough, rugged vehicle built for pulling power rather than speed. And power is essential for some of its tasks, such as pulling a 12-furrow plough through heavy soil. The usual type of tractor has huge wheels at the rear with a thick tread, and smaller wheels close together at the front. For extra grip on waterlogged ground crawler tracks are sometimes fitted instead.

Most tractors have a diesel engine, similar to those used in trucks and sometimes just as powerful (see page 97). Like a truck, the tractor usually has a wide range of gears. Generally, the engine drives only the rear wheels through an ordinary transmission system of clutch, gearbox, and

Above: Pigs eating from an electronic feeding machine in an intensive pig-rearing unit. A computer regulates the amount of feed according to the average weight of each batch of pigs.

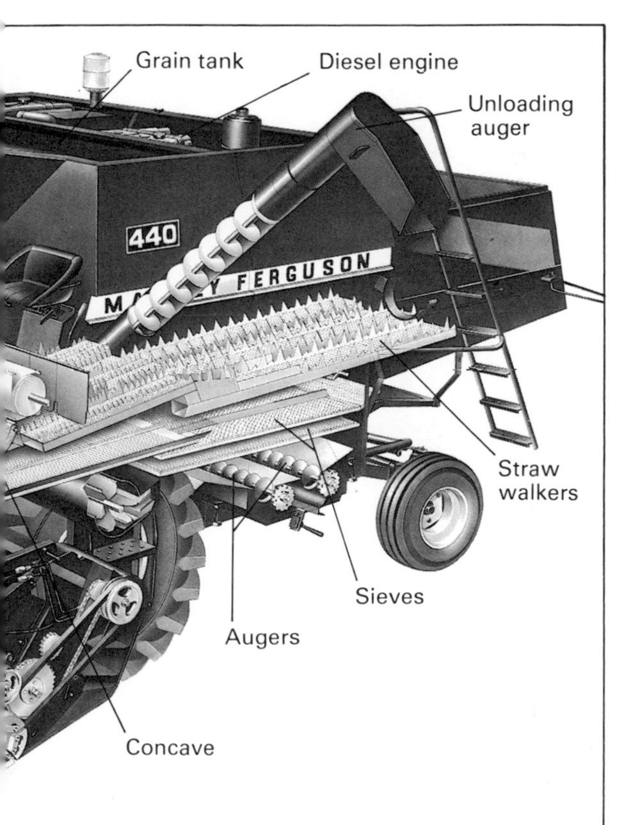

final drive. But the latest large tractors have four-wheel drive. All the wheels are driven to give extra pulling power.

The engine also drives another shaft that projects at the rear. This is the power-take-off (PTO) point, which provides power for implements that the tractor might pull. For example, it would drive the cutter bar of a forage harvester. The implement being pulled may be attached to a linkage that enables it to be lifted and adjusted by the driver. This is done by means of a hydraulic-ram system.

Ploughing

The crop farmer's main task in the autumn is preparing, or tilling, the soil to receive seeds or seedlings. If the soil is ill-prepared, the crops may fail. The farmer tills the soil in several stages, beginning with ploughing.

The invention of the plough over 5000 years ago, one of the great milestones in the history of civilization, led immediately to higher crop yields. The essential parts of a plough are the curved blades, called mouldboards, which have a sharp blade known as the share in the front. When the blades are pulled through the ground, they cut into and turn over a slice of soil, forming a distinctive pattern of ridges and furrows.

Modern ploughs may have as many as 12 blades, this being described as a 12-furrow plough. In the Middle Ages it took a man with a single-furrow plough pulled by oxen about two days to till a hectare of land. Today a 12-furrow plough, pulled by a tractor, can till more than 4 hectares in an hour.

Drilling

After ploughing, the farmer breaks up the ridges with various types of harrows. These are frames carrying spiked teeth or sets of sharp discs set at an angle. The soil may then be rolled, although this is usually done after the next stage – drilling.

The seed drill consists of a series of tubes that deliver the seed beneath the soil in straight lines and at the right depth. It incorporates a simple metering device to feed the seeds from the storage hopper into the tubes at a controlled rate. Fertilizer is often delivered to the soil with the seeds.

While the crop is growing, it is sprayed several times. Depending on the crop, it may be sprayed with insecticides, fungicides or herbicides to control, respectively, insects, fungus diseases or weeds. Then the time arrives when the crop is ready to be harvested. This is when the farmer prays for fine weather.

Whenever possible the farmer uses machines for harvesting. Mechanical harvesters are used these days not only for cereal crops such as wheat, but also for beans, root crops, cotton, apples and even soft fruits.

Combining

The most powerful harvester is the combine, used mainly for cereals. The combine harvester is so called because it combines the actions of reaping (cutting the crop) with threshing (removing the grain from the stalks).

Most combines are self-propelled and are powered by a diesel engine. In front they have a rotating reel above a cutter bar. Inside the cutter bar is a sharp-toothed knife blade moving back and forth. As the combine moves forward the blade cuts the grain stalks. The reel helps hold the stalks steady while they are being cut and guides them into a screw-like conveyor, or auger. In turn, the auger feeds the grain stalks into an elevator that lifts them into a threshing drum ("concave").

Rotating bars in the concave beat out the grain from the seed heads. Most of the grain and chaff fall onto a series

Below: The Fiat Series 80 tractor, one of the powerful machines now in use on modern farms. A six-cylinder diesel engine drives all four wheels. Four-wheel drive gives the tractor extra grip over difficult terrain. The Fiat 80 has a 16-speed gearbox and a two-speed power-take-off mechanism. The driver rides in comfort in an enclosed, well-insulated cab.

Above: Drilling spring wheat on an English farm. The modern seed drill works on the same principles as the original one invented by Jethro Tull in about 1700. The harrow behind the drill helps cover the seeds with soil.

Below: Even steep hillsides can now be put down to cereal crops, as on this French farm. The combine is fitted with an automatic levelling device, which keeps the main machine horizontal, while allowing the cutter to follow the slope.

of sieves ("riddles") that are shaken back and forth. A fan blows air through them to remove the chaff. Other augers feed the separated grain to an elevator that lifts it into a storage bin, which is periodically emptied. The straw meanwhile passes to the straw walkers. These also shake and transport the straw to the rear of the combine, where it is discharged, together with the chaff.

Machines For All Seasons

The combine is a complicated machine with a colossal output. Some of the big combines used in North America have a cutting width of up to 9 metres. They can harvest up to 4 hectares of land per hour.

The combine, however, has its drawbacks: it is very expensive (£40,000 or more); and for much of the year it is standing idle. The same is true for other farm machinery. So some manufacturers are now offering farmers a more flexible machine system. It consists of a main power unit rather like a tractor, together with a number of different bolt-on attachments. The attachments include a combine unit, for harvesting cereals and beans; a forage harvester, for cutting grass and other crops for SILAGE; and a corn sheller for harvesting maize.

Food Technology

Above: The blending control centre at a factory that manufactures margarines and a variety of cooking fats. The products are made by mixing together a number of different fats and oils. These are now obtained mostly from vegetable sources, such as groundnuts (peanuts), soya beans, coconuts, cottonseeds and sunflower seeds.

The technology of food production does not end at the harvest or the milking parlour. The various crops have to undergo a variety of processes before they reach us, the consumer. The methods of processing some farm products have changed little over the years. Grain is still ground into flour, which is made into bread. Modern technology has improved the speed at which these operations are carried out. Bread-making is now a highly automated, factory process. Similarly, the age-old processes of making butter and cheese from milk still continue, updated by modern technology.

New foodstuffs, however, now appear on the market alongside these traditional products. MARGARINE vies with butter as the spread for bread. "Meatless" meat is now found in stews and mince, in addition to real meat such as beef. These meat-substitutes, like the other products mentioned, are actually made from vegetable matter, usually soya beans. These beans are processed to give "meaty chunks" of what is called textured vegetable protein (TVP). Experiments are also well advanced on single-cell protein (SCP) derived from agricultural wastes and even petroleum. This is already being produced as animal feed.

Food Spoilage

One of the major branches of food technology is concerned with the preservation of food for transport and storage. The traditional process of cheesemaking, for example, can be considered a way of storing milk for long periods. Left by itself, milk keeps for only a day or so before it turns

Above: A pasteurization unit in a modern dairy, in which raw milk is heated briefly to kill any harmful bacteria present.

Below: A microwave oven and some of the foods that it can cook in minutes. It can bake a potato in about four minutes and roast beef in about 20. It cooks by means of microwaves, which penetrate right inside the food and agitate the molecules, causing the food to heat up quickly.

sour and becomes unfit for drinking. A freshly caught fish also will stay fresh only for a day or so and soon announces, by its smell, that it is decaying, or spoiling. The same goes for the meat of slaughtered livestock.

The action of bacteria, moulds and other microorganisms may cause foods to spoil. This not only makes them smell and taste bad, but it can also cause food poisoning. Chemical changes also cause spoilage. They may be brought about by the action of the oxygen in the air or by enzymes in the food itself.

Food Preservation

Even in prehistoric times, hunters found that they could preserve meat and fish by drying them. This process is effective because the microorganisms that cause spoilage cannot grow on dry food. Smoking is an extension of drying, and chemicals in the smoke greatly increase the storage life of the food. Other traditional methods of preserving foods include packing them in salt or sugar, and pickling them in brine or vinegar.

Such methods of food preservation are still used today. Drying, however, has been greatly improved so as to preserve the flavour, texture and appearance of the product. In the method of accelerated freeze drying (AFD) the food is frozen quickly in a vacuum. The water in the tissues rapidly freezes and then sublimes, or turns to vapour.

Freezing itself is now the second most important method of preservation. Although the Romans used ice to cold-store foods, freezing as a widespread preservation method had to await the invention of the refrigerator in the mid-1800s and of the quick-freezing process of Clarence Birdseye in the 1920s. Frozen foods need to be stored at temperatures below about −18°C otherwise they will deteriorate. This is the temperature of most domestic deep-freezes.

The most popular method of food preservation, however, is canning, in which food is packaged in sealed cans. The cans are heated for short periods in giant pressure cookers at temperatures up to about 130°C. The heat kills any organisms in the food, and the airtight cans ensure that no organisms can reach it afterwards. Brief heating is also the method used to preserve milk in dairies. Here the method is called PASTEURIZATION, after the Frenchman Louis Pasteur, who pioneered the investigation of the microorganisms that cause food and wines to spoil.

Most recent methods of preservation include treatment with antibiotics and irradiation with gamma-rays and ultraviolet rays. And the majority of processed foods contain chemicals of one sort or another to improve their keeping qualities. Other chemicals are added to improve flavour, notably the ubiquitous monosodium glutamate, which stimulates the taste buds.

MILACRON
MILACRON

Chapter 3

Industry

Practically all the goods we use today are made in factories. Factories take raw materials – chemicals, wood, minerals, ores – and transform them into products for use by us, the consumers, or by other factories.

The factory system grew up in the 1700s when manufacturers began to install machines in buildings and employ people to operate them. This transformed the making of goods from a home craft into an industry. It brought about an Industrial Revolution.

The Revolution still goes on, with machines taking over more and more of the work. The latest machines are robots working under computer control.

Decreasing numbers of people are now involved in the actual manufacture of goods. But increasing numbers are employed in the so-called service industries. These are concerned with marketing and selling products and with other activities, such as administration and insurance.

Robots at work on a Ford Sierra production line. They weld the body with inhuman precision hour after hour, unaffected by heat, glare and noise.

Raw Materials 1

Manufacturers use vast amounts of different materials to make the things they produce. They use such things as metals, plastics and woodpulp. These materials are made in turn from basic substances that we call raw materials. Metals are made from minerals taken from the ground. Most plastics are made from chemicals extracted from petroleum. Woodpulp is made from the wood from trees.

Minerals, petroleum and wood are important raw materials. Other vital raw materials are found all around us, in the air. The air is composed mainly of the gases oxygen and nitrogen. Both have many uses in industry. For example, nitrogen is made into ammonia, which is then made into fertilizers and many other products.

Mineral Ores

Thousands of different minerals are found in the Earth's crust. They are made up of combinations of two or more chemical elements – the building blocks of nature. The mineral bauxite is made up of the elements aluminium and oxygen. Chalk is made up of three elements – calcium, carbon and oxygen.

Bauxite is an example of an ore. An ore is a mineral from which a metal can be economically extracted. Aluminium can be economically extracted from bauxite. The minerals magnetite and haematite are important ores of iron. Copper pyrites and chalcocite are major ores of copper. Iron, aluminium and copper are the most important metals used in the modern world. Tin, lead, zinc, chromium

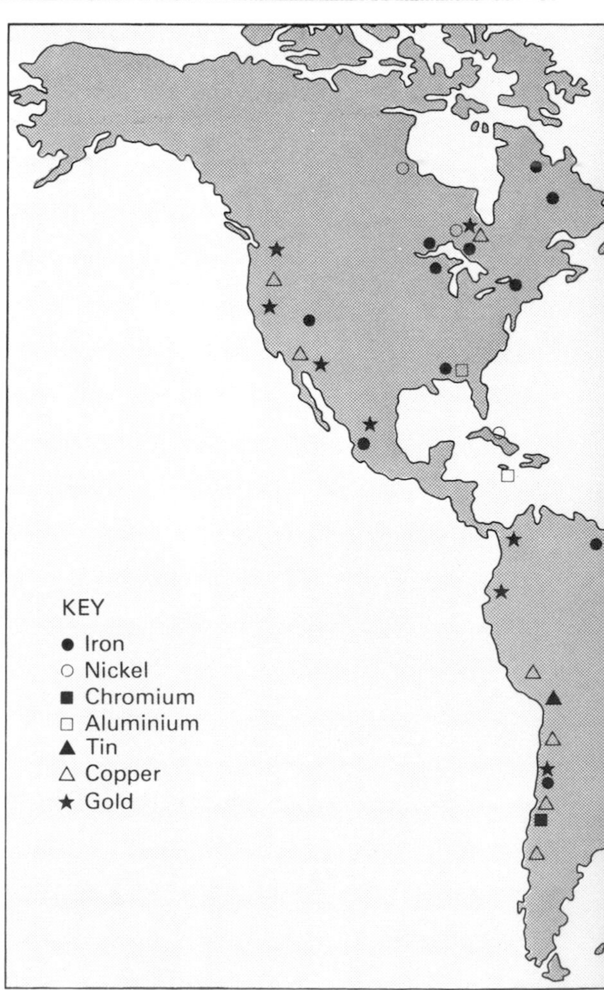

Above: A pile of logs stored at a pulp mill in Canada. Woodpulp is used to make rayon, cellophane and cellulose plastics.

Left: A huge walking dragline strips away the soil covering the coal deposit at an opencast mine. Coal will become an increasingly important raw material for chemicals in the future.

Below: A world map showing the location of major deposits of metal ores.

and nickel are also widely used. All of these metals are extracted from ores.

Only a few metals can be found in metal form in the ground. They include gold, silver and platinum. These are known as precious metals because they are rare, have an attractive appearance and do not corrode, or "rust" away.

Chemicals from the Rocks

Industry requires many other minerals besides the ores to provide the chemicals to make its products. The fertilizer industry uses millions of tonnes of phosphate rock. It treats the rock with sulphuric acid to make the widely used fertilizer superphosphate.

Practically all industries, in fact, use sulphuric acid in one way or another. It is the most important of all chemicals, and is often called the "lifeblood" of industry. It is made from the minerals sulphur and iron pyrites.

Common salt is another vital raw material, which can be made into caustic soda, baking soda, glass, chlorine and hydrochloric acid. It can be obtained from the rocks as rock salt, or from the seas. Seawater is a solution of common salt and other chemicals as well. Even the sand on the seashore is a valuable raw material.

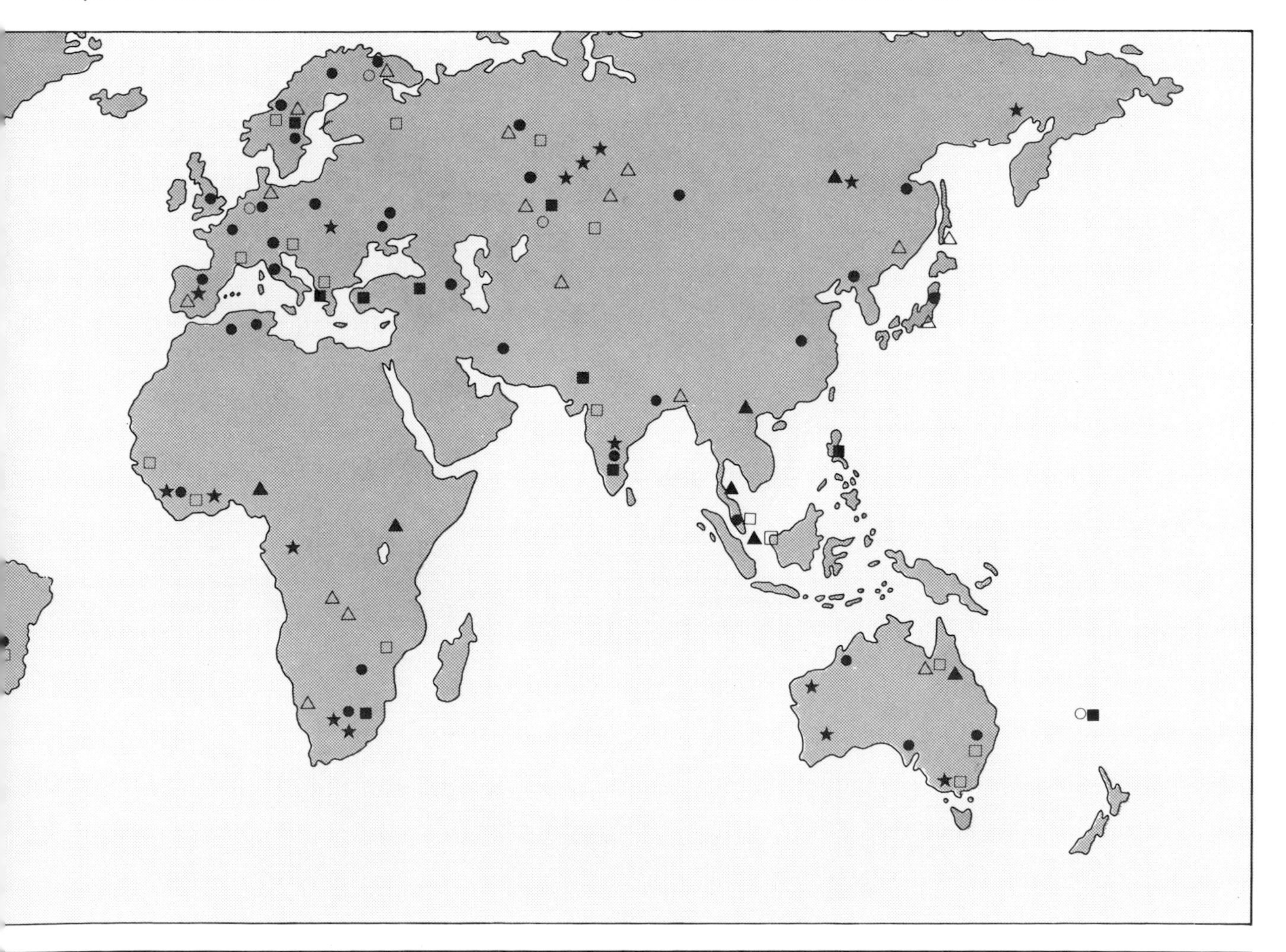

Raw Materials 2

Chemicals from Coal and Oil

The Earth has a particularly rich source of chemicals in coal and petroleum (crude oil), although they are better known as fuels. Coal and petroleum contain a host of organic chemicals. They are so called because they are the remains of once living things (organisms). They are made up mainly of hydrocarbons – compounds of carbon and hydrogen.

In oil refineries, for example, the hydrocarbons in petroleum are separated out and converted into a large variety of different chemicals (see page 64). Industries use these chemicals to make such things as plastics, dyes, drugs, paints and synthetic fibres and rubbers. Coal is not such an important source of organic chemicals as it once was. But it is likely to become so again when the oil wells run dry.

Above: Bucket-wheel excavators at an iron-ore mine in Port Hedland, Western Australia. The wheel digs into the ground as it rotates, depositing the soil on a conveyor which carries it to the rear.

Future Supplies

Basic supplies of petroleum are running out, as are deposits of other minerals. To satisfy the demands of modern civilization, unbelievably large amounts of minerals have to be mined. In the United States and Russia alone, 300 million tonnes of iron ore are taken from the ground each year.

By the beginning of next century there could be a

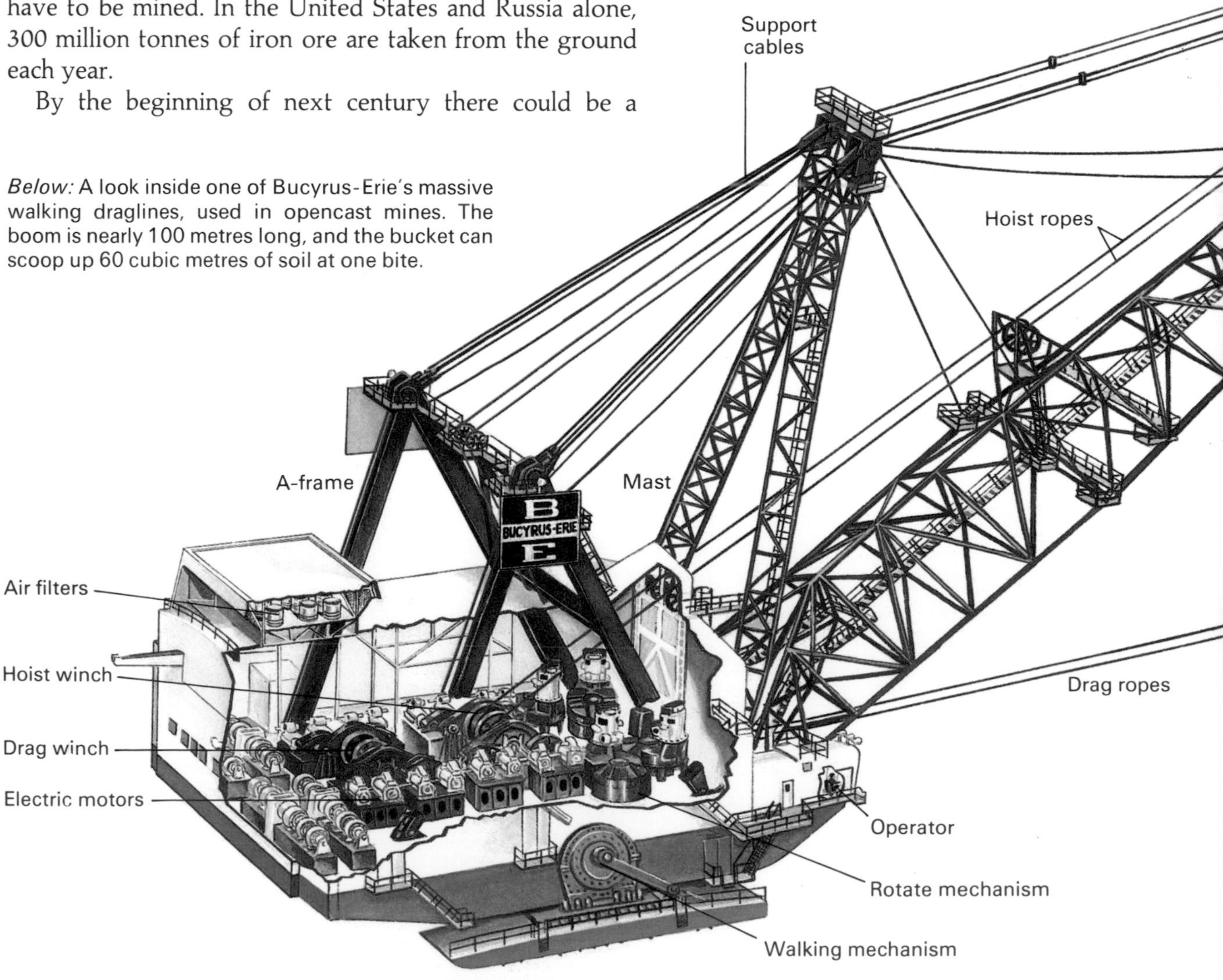

Below: A look inside one of Bucyrus-Erie's massive walking draglines, used in opencast mines. The boom is nearly 100 metres long, and the bucket can scoop up 60 cubic metres of soil at one bite.

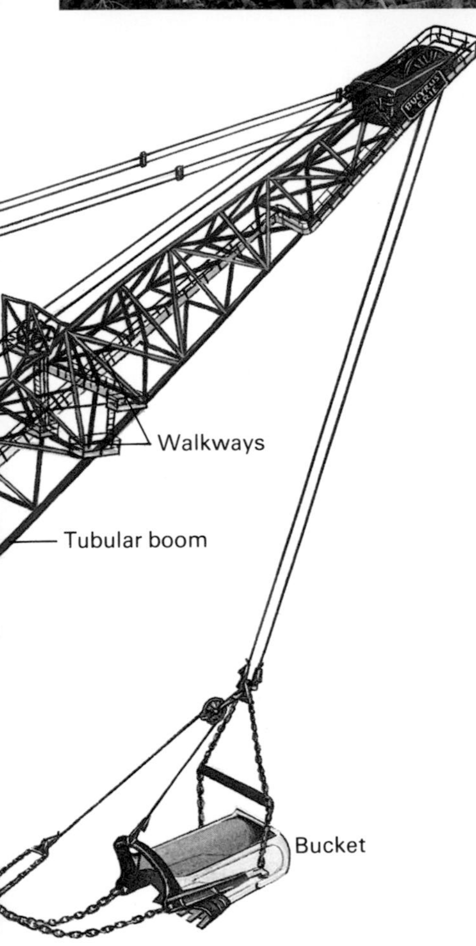

Top: A collection point for aluminium cans in the Florida Keys. The cans are eventually smelted down and the metal recycled.

Right: A Landsat picture of the Mississippi River. Landsat is an Earth-survey satellite that provides false-colour pictures. From them, geologists can often identify hitherto unknown deposits of valuable minerals.

serious shortage of many vital metals, including copper, gold, silver, platinum, tin, zinc, lead and uranium. But other sources of these metals may by then be being tapped. One such source is the oceans, which contain very many metals in the form of compounds dissolved in water. Already one metal – magnesium – is successfully obtained by treating seawater.

On the deep ocean floor are found mineral lumps known as manganese nodules. They contain particularly large quantities of manganese, copper and nickel. Experiments in mining these nodules, by dredging, are already under way.

Promising new deposits of minerals are also being located from space. Earth survey satellites like Landsat take pictures of the Earth not only in visible light but also in invisible light (such as infrared). The pictures are then printed in false colours (see below) which show up unsuspected features in the rocks. These features sometimes indicate new mineral deposits.

Recycling

But no matter what new sources of raw materials are found, they will one day run out. And more and more it is becoming sensible to try to re-use the materials wherever possible. This is the idea behind recycling. As yet, recycling is practised on only a small scale, mainly with paper, glass and cans. But in the United States and Britain alone each year, some 35 million aluminium cans are still thrown away. The loss of this amount of aluminium will become quite unacceptable in the years to come.

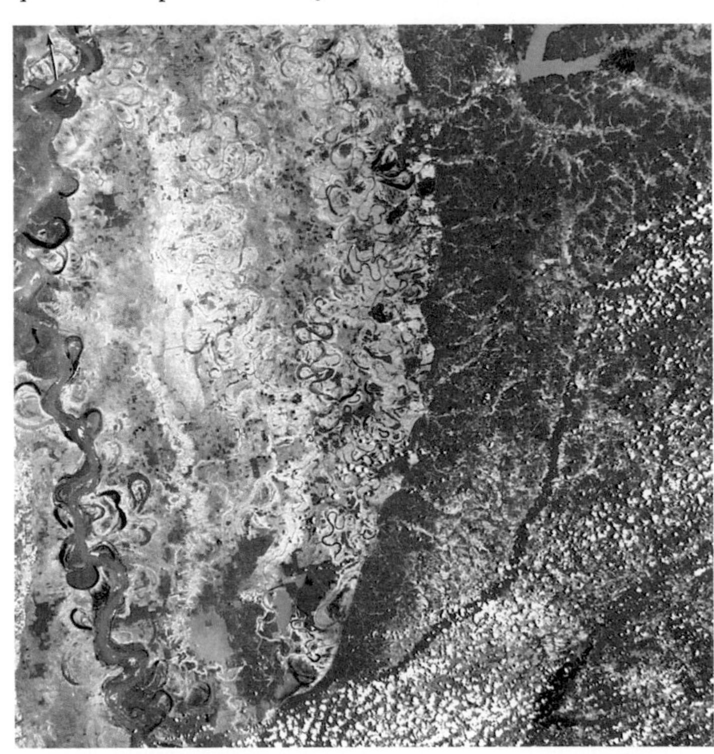

Extracting Metals 1

Of all the materials we use, metals are by far the most important. They are in general strong, hard and easy to shape. Without them, we could not build engines, machines or giant structures such as bridges and skyscrapers. Metals also conduct electricity, which makes it possible to build electric generators to produce electricity, and electric motors to power machines. Without machines and electrical power, our modern way of life would be impossible.

The progress of people along the path of civilization has depended largely on discoveries in the field of metallurgy – the extraction, processing and application of metals. Until about 6000 years ago people lived in a Stone Age. They made their tools and weapons from stone, usually flint, and to a lesser extent from wood and bone. But these materials are not very strong and break easily. Occasionally, people found lumps of native metals, such as gold and copper. They soon realized that these were far better materials for making tools. But native metals are very scarce.

Above: A bronze statue of George Washington in Boston, Massachusetts. Bronze has been a favourite material for casting for 5000 years and is still one of our most valuable alloys. It is easy to melt, flows easily when molten, and resists corrosion.

Smelting

Then in about 4000 BC, somewhere in the Middle East, a fortunate accident occurred. Someone happened to build a fire on a deposit of copper ore. The hot, burning wood attacked the copper ore, and changed it into copper metal. This was a great step forward; people had discovered how to smelt metals.

About 500 years later, they discovered how to smelt mixtures of copper and tin ores into bronze, a metal even more useful than copper. This period of their history is now known as the Bronze Age. About 1500 BC they found out how to smelt iron ores into iron, and they began an Iron Age. We are still in an Iron Age today, for iron is by far the most important of all metals.

Alloys

Iron is itself not a particularly strong metal. But because it can be made into steel it has many uses in the modern world. Steel is a mixture, or alloy, of iron and carbon, and usually other metals as well. It is much stronger and harder than iron itself.

Most metals are in fact used in alloys with other metals. Bronze is an alloy of copper and tin. Brass is an alloy of copper and zinc. Stainless steel is an alloy of iron, chromium and nickel. By carefully selecting the alloying metals – choosing the right "recipe" – scientists can produce alloys with special properties.

Ordinary steel, for example, is strong, but it rusts easily. Chromium and nickel are rust-resistant. When combined with steel, the result is stainless steel.

Some of the most advanced alloys are used in jet engines, where they must remain strong at temperatures

Below: The electrolytic cell used for the production of aluminium. Electricity is passed through a molten solution of alumina and cryolite. The aluminium collects on the floor of the cell, on the cathode.

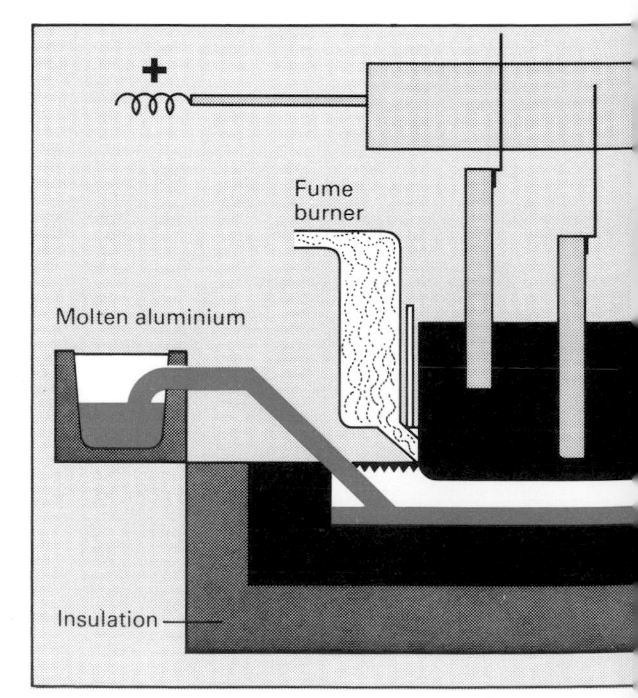

Above: A Rolls-Royce RB-211 turbofan jet engine. It is constructed from a large number of special alloys, able to withstand stresses at high temperatures. These are made from such metals as nickel, chromium, cobalt, titanium and tungsten.

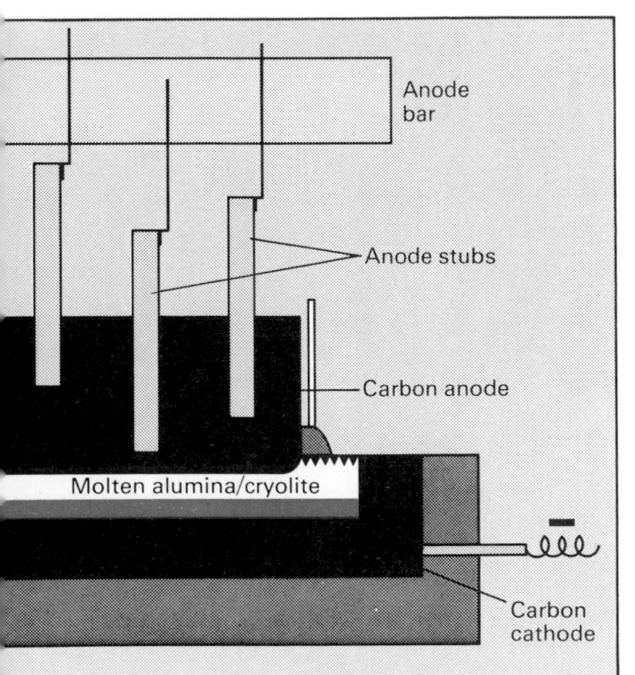

of over 1000°C. These are known as superalloys and contain metals with high melting points, such as titanium and tungsten. Sometimes in rocket engines materials called CERMETS are used, which are mixtures of metals with ceramic materials.

Electrical Refining

Some metals cannot be produced in the same way as iron. For example, aluminium cannot be smelted in a blast furnace. But it can be extracted from its ores by electricity. The process works because certain chemical compounds can be split up into their elements by passing electricity through them. This is known as ELECTROLYSIS.

In aluminium production (the Hall–Héroult process) electricity is passed through a molten mixture of alumina (aluminium oxide) and another mineral called cryolite. The alumina is obtained from the mineral bauxite, which is the only ore of aluminium. As the electricity goes through it, the alumina splits up into oxygen and aluminium metal.

Electrolysis is also used in extracting and refining other metals – for example, copper. Copper is obtained from some of its ores by treating them with sulphuric acid. The result is a solution of copper sulphate. Copper is obtained by passing electricity through this solution.

Extracting Metals 2

Above: Steel waste at a scrapyard in Florida. Steel scrap is usually added to molten pig iron in the steelmaking furnaces. In electric furnaces only steel scrap is used.

Above: An electric-arc furnace being tapped. The white-hot metal is poured into a travelling ladle. It will then be made into ingots by casting in moulds, or directly into slabs by continuous casting.

Below: A blast furnace and associated plant. Hot air is blasted into the furnace, making the coke burn fiercely. The iron ore is reduced to molten iron metal, which trickles down to the hearth. The furnace gases are cleaned and then brought back to heat the stoves, which preheat the air going into the furnace.

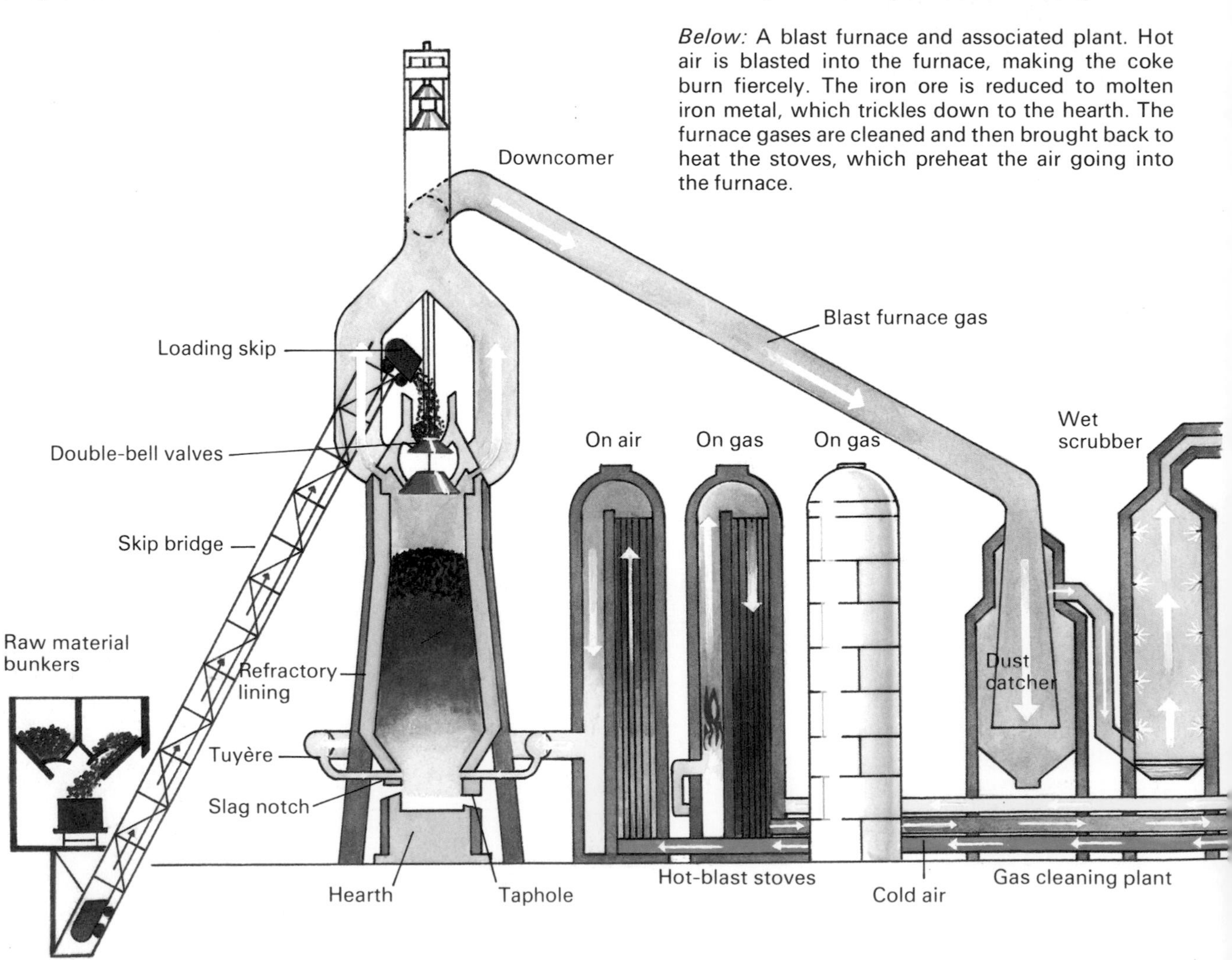

The Blast Furnace

Smelting is still the main method of producing many metals, including iron, lead, zinc and copper. Iron is produced in huge blast furnaces up to 60 metres high. These furnaces work continuously for weeks at a time, producing thousands of tonnes of iron every day.

The blast furnace is a tower, lined with heat-resistant bricks. Iron ore is charged into the furnace, together with limestone and coke. Hot air is then blasted through the furnace. The coke burns fiercely and raises the temperature in the furnace to more than 1500°C. It combines with oxygen in the iron ore, and reduces this to iron metal. The metal is molten and trickles down to the bottom of the furnace.

Impurities in the ore pass into the limestone and form what is called a slag. The slag is also molten and collects on top of the iron. From time to time the furnace is tapped – the molten iron is run out either into moulds or into huge buckets called ladles.

Making Steel

The iron that comes from the blast furnace is known as pig iron. It is very impure and needs to be purified, or refined, before it becomes really useful. Some is slightly purified into cast iron, which is widely used for making such things as car engine blocks by casting (see page 48). Most pig iron, however, is refined into steel.

The most important steel-making method today is the basic-oxygen process. This is outlined in the diagram (right). In the process molten pig iron is poured into a converter, a conical container made of steel and lined with heat-resistant bricks. Pure oxygen is then blown into the molten iron through a lance. In a spectacular "fireworks" display, the impurities in the iron either burn out or form a slag.

The basic-oxygen process is very quick. It takes only about 45 minutes to convert a charge of about 350 tonnes of pig iron into steel. This is why it has largely replaced the old open-hearth process, which used to be the main steelmaking method. That process took up to ten hours to make the steel.

The best quality steel is made in electric-arc furnaces. It is made not from molten pig iron, but from steel scrap. The scrap is melted in the furnace by means of an electric arc – a kind of continuous spark – from huge carbon rods. Lime and iron ore are added during the process to make a slag to absorb impurities.

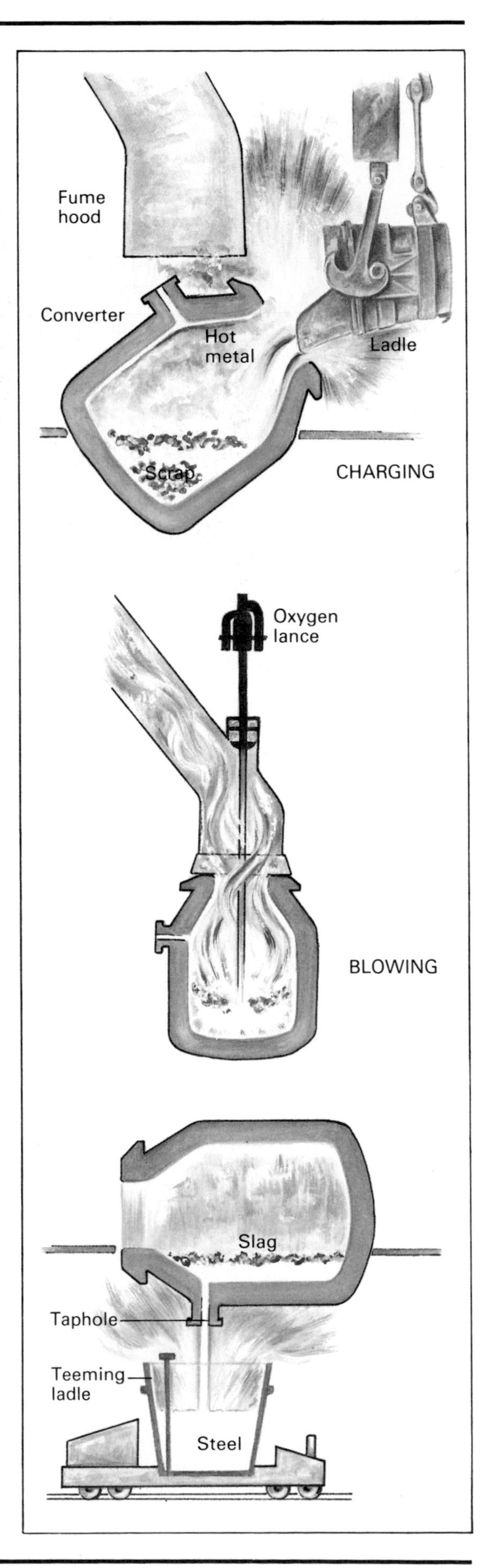

Right: Making steel by the basic-oxygen process. Molten pig iron is charged into the tilting furnace, or converter (top). Pure oxygen is then blown into the molten metal (middle). When the impurities have burned out, the furnace is tapped (bottom).

Shaping Metals 1

Above: A laser slices through metal like a knife through butter.

Opposite: An engineer checks the dimensions of a giant turbine shaft, which has just been machined on a lathe.

Below: Outline of the continuous-casting process.

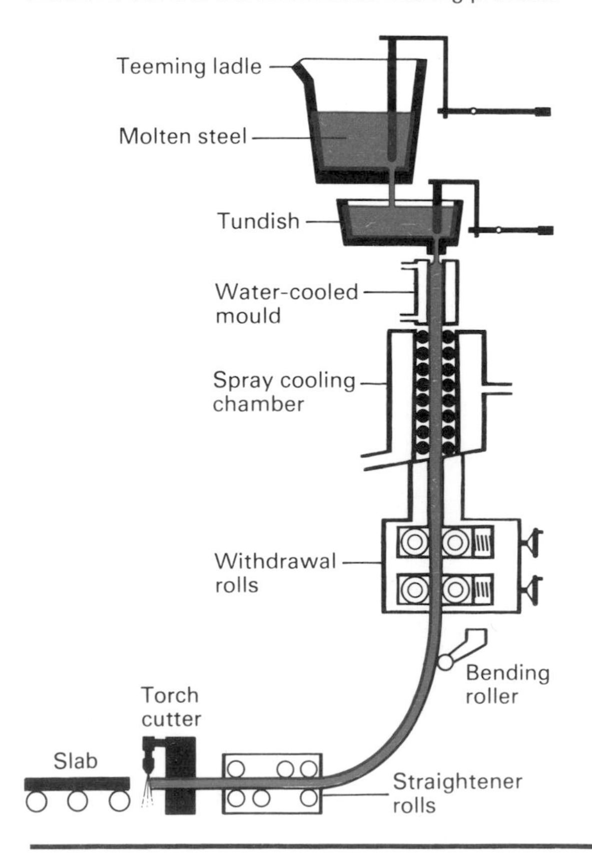

Metals may be shaped in a variety of ways. Everything depends on the metal and what it is to be used for. A few metals, including gold and copper, can be shaped easily when cold by hammering, for example. Gold can be hammered, without cracking, into a sheet so fine that it becomes transparent!

Most metals, however, are too hard or too brittle to be shaped when cold. So they are shaped hot, either when they are molten or when they are red-hot. The commonest shaping methods are casting, rolling and forging and machining, which are described below. Other common methods include drawing and extrusion. In drawing, a rod of metal is pulled through a hole in a die. In extrusion, hot metal is shaped by being forced through a hole in a die.

Metallurgists have had to devise novel methods of shaping some of the products they want to make. They often shape metals with high melting points, such as tungsten, by powder metallurgy. In this method powdered metal is pressed into shape in a mould and then heated strongly. Explosives may be used to force metal into shape. Sparks and chemicals may be used to "eat away" metal to the correct shape. Other methods may use lasers, electron beams and even powerful magnets.

Casting

This was the first method used on a large scale to shape metals, beginning with copper and bronze. And bronze is still cast today, into statues and ships' propellers, for example. Casting takes place in a workshop called a foundry. It is a process in which molten metal is poured into a mould and allowed to cool. When cold, the metal takes the shape of the mould, like a jelly does when it sets.

Most moulds are made from a special mixture of sand and clay. A model of the object to be cast is placed in a box and the sand mixture is packed firmly around it. When the mould has hardened, the model is removed, leaving a cavity of the shape desired.

Two holes are made in the top of the mould. The molten metal is poured through one, while the other allows the air inside to escape. When the metal has cooled, the mould is broken up and the cast object removed.

In some industries permanent metal moulds, or dies, are used. Many toys are made by diecasting on automatic machines. In this process hot metal is injected into a closed, water-cooled die and immediately freezes. The die then opens and the diecast object is ejected.

Casting is often a preliminary process before other shaping methods, such as rolling. Molten steel, for example, may be cast into large moulds called ingots, or formed into more convenient shapes by continuous casting. In this method, molten metal is poured into an open-ended mould, cooled by water. It emerges as a red-hot, solid slab.

Shaping Metals 2

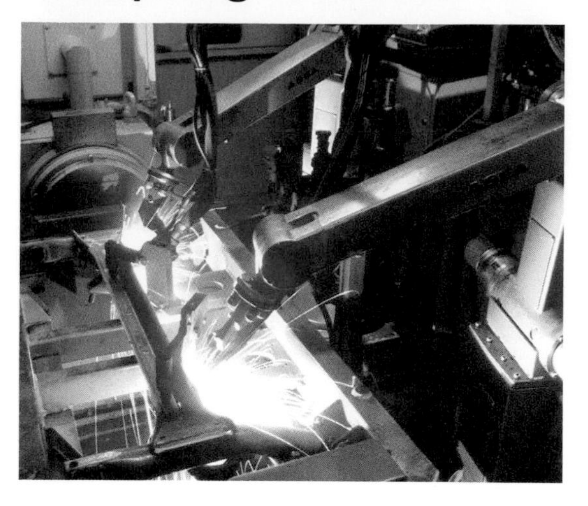

Above: Automatic electric welding of car body parts by robot. Welding is the process of joining pieces of metal together by fusion. The pieces are brought into contact and the joint between them is heated until the metal becomes soft and fuses together. Sometimes extra metal is added to the joint from a filler rod. When the metal cools down, the pieces are firmly bonded.

Below: An early stage in the forging of a turbine shaft from a red-hot steel ingot on a hydraulic press. The ingot is squeezed into shape by gradual, but unrelenting pressure from a hydraulic ram. After forging, the shaft will be machined and will end up looking rather like the shaft shown on page 49.

Rolling

A great deal of metal is used as plate (as in shipbuilding) and sheet (as in making car bodies and cans). Both these products start out as cast ingots or slabs, which then have to be reduced in thickness by rolling. In the rolling mills the metal slabs are first heated red-hot in a furnace and then passed back and forth through a series of heavy rollers, which are rather like old-fashioned mangles. Each set, or stand of rollers, reduces the thickness of the slab a little each time. As the thickness of the slab is reduced, its length increases.

A slab may go slowly into the first set of rollers measuring 10 metres long and 250 millimetres thick. It may emerge from the last set of rollers over 1500 metres long and 2 millimetres thick, and travelling at a speed of 100 km an hour!

Rolling mills produce not only plate and sheet metal, but also other shapes as well. They produced shapes like girders for bridges and rails for railway track. To do this they use shaped rollers.

Forging

When people first made iron, they could not cast it into shape because their furnaces were not hot enough to melt it. So they hammered the spongy red-hot metal into shape instead. This method of shaping is called forging. Hand forging is still practised today by the blacksmith to make horseshoes and ornamental ironwork.

In industry, however, machines are now used to do the hammering. A common one is the drop forge, which has a heavy ram that falls onto the metal. The ram is raised by air or steam pressure. The most powerful forges, however, do not shape by means of a sudden hammer blow, but rather by a gradual squeezing action. They are known as forging presses and work by hydraulic (liquid) pressure. Some can exert a pressure of 10,000 tonnes or more. They can squeeze a massive steel ingot as though it were a piece of putty.

Similarly, hydraulic machines are used to shape cold metal sheet, as for car body parts. The process is then called pressing. Also, machines similar but smaller than the drop forge are used to shape cold metal in a process called stamping. Coins and medals are made by stamping.

Machining

Most metal parts, no matter how they are shaped, need further attention before they are ready for use. For example, a cast-iron engine block has to be drilled to take bolts. Threads must be cut in a metal rod to make it into a bolt. Drilling and thread-cutting machines are examples of what are called machine tools.

Machine tools are power-driven machines designed to

Left: The well-named oil production rig Magnus, after construction. It was made by welding together many kilometres of steel tubing. It is shown on its side, and its size can be gauged by comparison with the double-decker bus parked alongside.

remove metal from an object, or workpiece. They need powerful electric motors to force cutting tools into the metal. The tools have to be made of hard steels containing tungsten and chromium, otherwise they become blunt quickly.

The most useful machine tool is the lathe, which is found in practically all engineering workshops. It carries out a machining operation called turning. The workpiece is held and rotated, while cutting tools are moved into it. Drilling is another routine operation, in which holes are bored into a workpiece by means of a rotating drill bit. A third common machine tool is the milling machine. This cuts metal with a rotating toothed cutting wheel, while the workpiece moves beneath it. Other machine tools remove metal by planing and grinding.

Accurate machining is essential in all mass production because of the need to produce nearly identical parts each time. Many machine tools now work automatically under computer control and play a vital role in factory automation (see page 65).

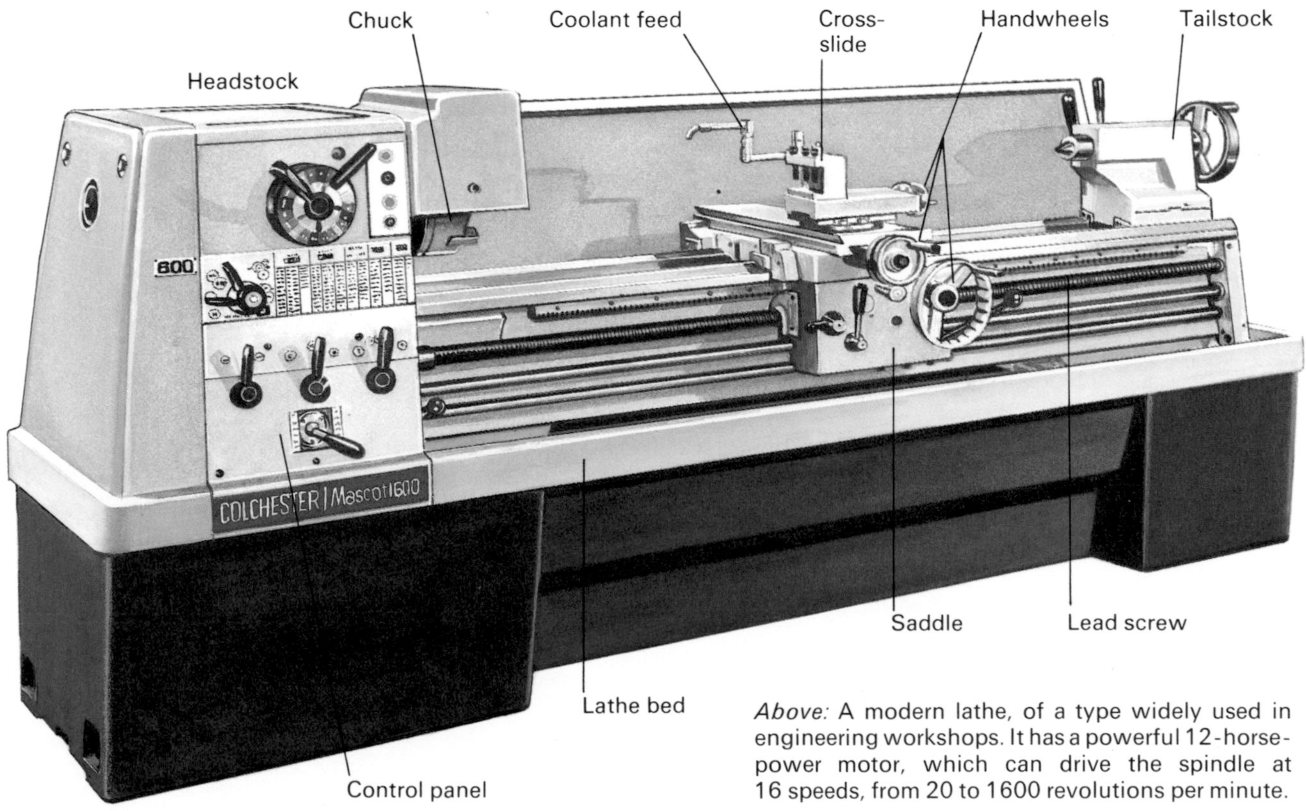

Above: A modern lathe, of a type widely used in engineering workshops. It has a powerful 12-horse-power motor, which can drive the spindle at 16 speeds, from 20 to 1600 revolutions per minute.

Textiles

The art .of spinning fibres into yarn and weaving yarn into cloth has been practised for at least 10,000 years. But until about 200 years ago spinning and weaving were done at home. Then came the invention of a number of spinning and weaving machines that turned textile-making into the first great industry.

The machinery for spinning and weaving has improved steadily over the years. But the greatest revolution in textile-making has been brought about by the introduction of a host of new fibres. Until about a century ago, the only fibres available came from plants and animals – cotton, from the boll of the cotton plant; linen, from the stalks of the flax plant; wool, from the fleece of the sheep; silk, from the cocoon of the silkworm. Then came the discovery of man-made fibres, such as rayon and the synthetic fibres like nylon.

Rayon

Rayon is the man-made fibre produced in the largest quantities today. It was once called artificial silk. It is made from the cellulose in woodpulp and cotton waste. It is called a regenerated fibre, because the cellulose is first dissolved and then reformed, or regenerated. The most important type of rayon is called viscose. It is made by dissolving cellulose material in caustic soda and carbon disulphide. The syrupy solution formed is then pumped through the holes of a device called a spinneret into an acid bath. In contact with the acid, the streams of solution coming from the holes change into solid threads, or filaments, of pure cellulose, which are then wound onto a bobbin. This process is called wet spinning. Other cellulose fibres called acetate and triacetate are made by treating cellulose material with acetic and sulphuric acids.

The long filaments may be used in this form, or they may be chopped up into short lengths, as staple fibres. These fibres can then be spun in the usual way, either by themselves or in mixtures with natural fibres.

Synthetic Fibres

Most types of man-made fibres, however, are not made from natural materials like cellulose, but from chemicals. They are synthetic. They are kinds of plastics that can be drawn out into fine threads (see page 63). Most synthetic

Top right: Continuous filaments emerging from the acid bath, during rayon production by wet spinning.

Right: Spinning thread on a rotor-spinning machine. Loose fibre slivers (bottom) pass up into the spinning units, where they are compacted into thread by a rotor rotating at 70,000 revolutions per minute. The emerging threads are then wound onto bobbins (top).

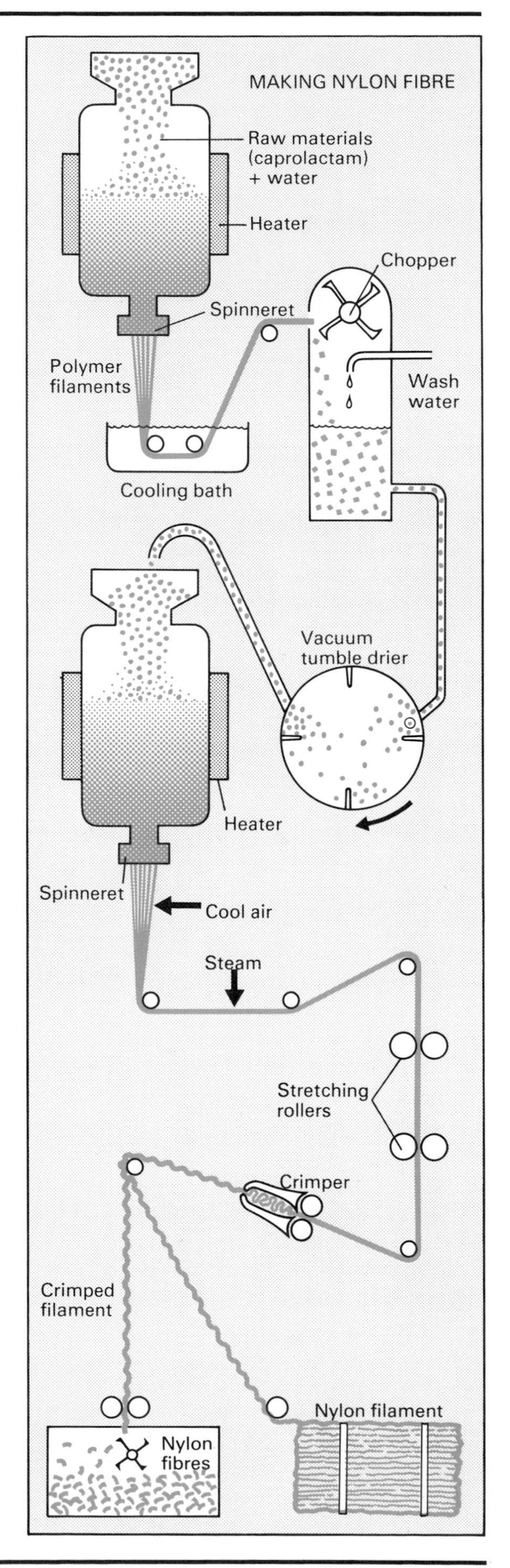

Right: Stages in the production of synthetic nylon fibre. It is made from a petrochemical called caprolactam. This is first polymerized into plastic and then turned into fibre by melt spinning.

fibres are made from petroleum chemicals, produced at oil refineries.

The best-known and original synthetic fibre is nylon. To make fibres, nylon chips are melted and forced through a spinneret into cold air. The streams of molten liquid that emerge solidify into long filaments. This process is called melt spinning. Polyester fibres such as Terylene and Dacron are produced in a similar way. Acrylic fibres such as Acrilan and Dralon must be prepared in a different way because they decompose when they are heated. They are often prepared by wet spinning as for rayon; or by dry spinning, being dissolved in a solvent that evaporates in the air.

Man-made fibres have many desirable properties that often make them superior to the natural fibres. Viscose is strong and easily dyed. Acetate fabrics have the look and feel of silk, keep their shape well and can be permanently pleated. Synthetic fibres are exceptionally strong, resist insect attack and rotting, and do not absorb water. This last property makes them easy to wash and "drip dry" quickly.

In addition, each fibre has special qualities that make it suitable for a particular use. Nylon is slightly elastic, making it useful for stockings and mountaineering ropes. Polyesters on the other hand resist stretching, and therefore fabrics made from them keep their shape well. Acrylic fibres make up into exceptionally soft and warm fabrics.

Modern Textile Machines

Spinning machines take loose ropes, or slivers, of fibres, draw them out into finer thread, and impart a twist to make them into a firm yarn. One of the most widely used machines is the ring-spinning frame. On this machine, the yarn is guided onto the bobbin and twisted by a clip ("traveller") that moves around it on a ring. The latest frames carry as many as 500 bobbins, which rotate at a speed of 12,000 revolutions per minute or more. The recently introduced rotor-spinning machines can work even faster, and produce over 120 metres of yarn a minute.

The modern looms used for weaving work on much the same principles as the simple hand loom that has been used for centuries, but at an incredible speed. The latest ones do not have an ordinary shuttle to carry the weft (crosswise) thread. They pass the weft by means of a projectile, a rapier-like rod, or even jets of air or water. Some machines can put down over 400 metres of weft every minute.

Ceramics

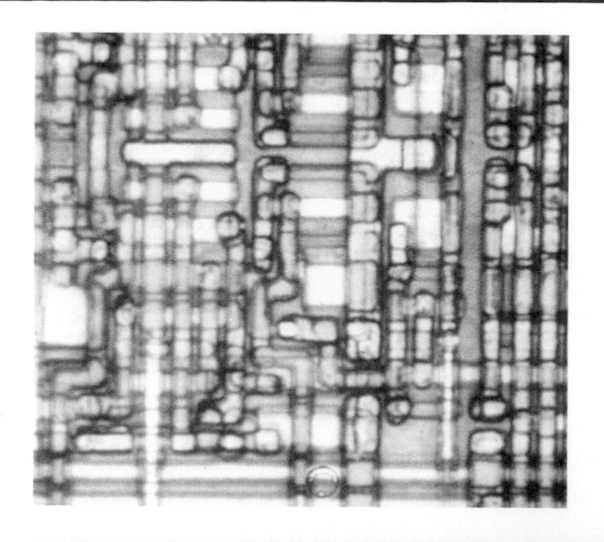

Clay is one of the most plentiful substances on Earth. It is also one of the most useful because it can easily be turned into a variety of different products, such as pottery, bricks and tiles. All these products are made by shaping moist clay and then baking it in an oven, or kiln. The clay sets hard and becomes strong and hardwearing. We call baked clay products, ceramics.

We also use the term "ceramics" to include other products made by baking or burning other earthy materials in a kiln or furnace. Cement, the essential ingredient of CONCRETE, is a ceramic material. So are glass and the heat-resistant materials we call refractories.

Pottery

The ordinary kind of pottery, used for everyday crockery, is earthenware, which is fired (baked) at a fairly low temperature (about 1000°C). It is by itself dull and porous, and needs to be glazed to make it waterproof. A stronger and waterproof pottery is stoneware, which is fired at temperatures up to about 1400°C. One of its main uses is for sewerage piping. The finest pottery is porcelain, which is waterproof, glass-like and translucent – it lets light through. The ingredient in porcelain that makes it a superior product is pure white china clay, or kaolin. Bone china is an imitation porcelain that contains bone ash.

Glass

Glass has been in use since the days of ancient Egypt, and it is still one of the most versatile materials available. It is cheap and easy to shape. It is waterproof, transparent and easy to clean. No common chemicals affect it. The ordinary glass used to make windows, jars and bottles is called sodalime glass. It is made by heating a mixture of sand, soda ash and limestone in a furnace to a temperature of about 1500°C. The mixture melts, or fuses, to give a red-hot liquid. As the liquid cools, it gets thicker until it sets hard and becomes transparent.

Different types of glass can be obtained by varying the recipe. By including lead oxide, lead or crystal glass is obtained, which is noted for its diamond-like sparkle and brilliance. Glass containing large amounts of lead is used in the atomic-energy industry because it can block dangerous radiation. Another common type of glass, borosilicate glass, contains boron compounds. Pyrex is a well-known trade name for such glass, which is tough and resistant to heat. It is widely used for making cooking ware and laboratory apparatus.

Many methods are used to shape glass, including blowing, moulding and casting. Ordinary sheet window glass is made by drawing up a sheet of molten glass vertically upwards from a red-hot furnace. Better quality flat glass called float glass, is made by floating molten

Left: The circuits in a silicon chip, photographed through a microscope. The raw material for the chip, silicon, is a ceramic material. It is made by heating silica (pure quartz) with carbon in a high-temperature electric furnace.

Right: Inside the float bath in the float-glass process. Red-hot molten glass floats perfectly flat on the surface of the molten tin.

Below: The refractory material from which the space shuttle tiles are made. Its remarkable insulating properties can be judged from this photograph. It can be held with bare hands when the outside is cool but the inside is still red-hot!

glass on a bath of liquid tin. The safety glass used for the windscreens of vehicles and aircraft may be toughened or laminated. Toughened glass is made by suddenly cooling a hot glass sheet with a blast of cold air. Laminated glass consists of a sandwich of two sheets of glass with a film of clear plastic in between. Both types of glass do not form sharp splinters when they shatter.

Glass fibre is made by forcing molten glass through a network of tiny holes. It is used as reinforcement for plastics, in the material known as glass-reinforced plastic (GRP), popularly called fibreglass. Very pure glass fibres have exciting uses in the field of FIBRE OPTICS.

Refractories

These are materials with excellent resistance to heat. They are most widely used to line furnaces. Most furnaces are lined with bricks made of naturally occurring minerals such as silica, dolomite and magnesite. For higher temperature resistance a variety of oxide materials are used, such as alumina, zirconia and beryllia, all of which have a melting point about 2000°C. For certain applications, pure carbon can be used, which can resist temperatures over 3000°C.

Some of the best refractories are made from carbides – compounds of heat-resistant metals and carbon. An example is tungsten carbide, a material used to make cutting tools that remain sharp even when red-hot. Many special-purpose refractories have been developed in the aerospace industry for use in the high-temperature parts of jet and rocket engines. They are called cermets, and are combinations of ceramics and metals. They combine the heat-resistance of ceramics with the strength of metals. They contain such materials as tungsten and titanium carbides and powdered iron and cobalt. One of the most interesting refractory materials is that used for the heat-shield tiles of the space shuttle. Made from silica fibres, it is one of the best insulating materials known.

Oil Refining

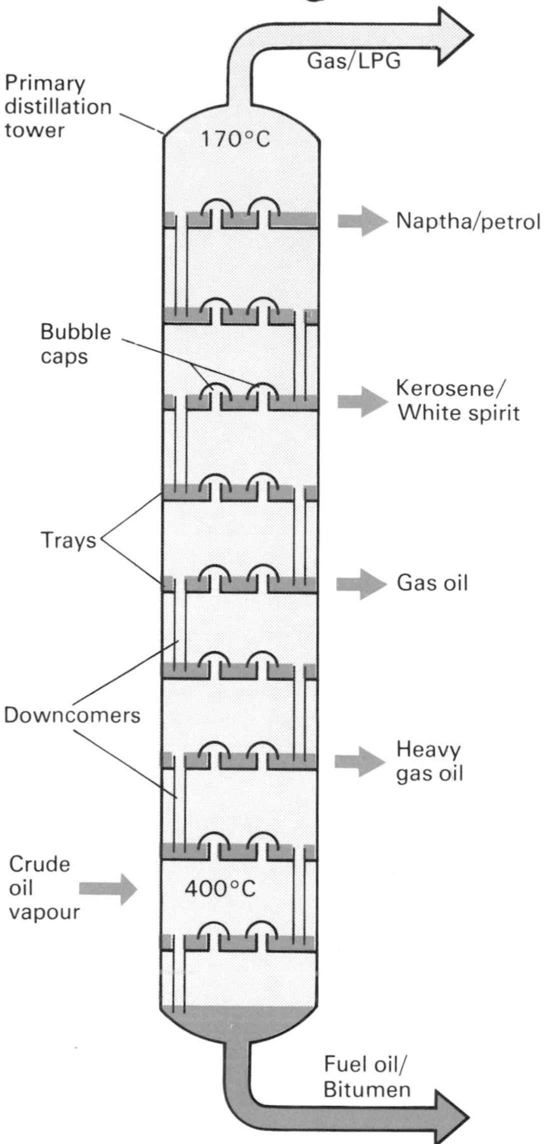

Above: A simplified diagram of a fractionating tower, in which the first refinery operation of distillation is carried out. Crude oil vapour is introduced near the foot of the tower and rises upwards. The rising vapour mixes thoroughly with the liquid in the trays as it bubbles through the bubble caps (or similar devices). The various fractions in the vapour condense into liquid in the trays at different levels. In practice the fractionating towers are up to 80 metres tall and have about 40 trays.

Top right: The catalytic cracker at the Europoort refinery near Rotterdam. Catalytic cracking breaks down heavier oils into lighter, more useful products, such as petrol.

Right: This flow diagram shows the operations that occur in one small section of a refinery. It shows the processes involved in producing ethylene and other chemicals from gas oil.

One of the most important of all chemical plants is the oil refinery. At an oil refinery, petroleum, or crude oil, is converted into a wide variety of fuels and chemicals from which many different products are made.

Crude oil is a mixture of substances called HYDROCARBONS. The first stage of refining is to split up this mixture into parts (fractions) that boil at different temperatures. This is done by DISTILLATION. The oil is heated in a furnace until it becomes a vapour. The vapour is then led into a tall column and up through trays held at different temperatures, with the coolest at the top. The various fractions condense, or change back into liquid, at the appropriate temperature level.

From the top level petrol is produced. At lower levels come kerosene (or paraffin), diesel and heating oils, fuel oil and thicker lubricating oils. Near the bottom are heavy oils and tarry bitumen. Petrol, kerosene and diesel and fuel oils are our most important fuels for cars, jet planes, lorries and ships, respectively.

Useful substances are found in the gases that come from the top of the distillation column. They include propane and butane, which can be easily turned into liquid and used as "bottled gas".

Improving the Product

The heavy oil is not much use by itself, but chemists have found how they can change it into petrol. They do so by a process called cracking. The heavy oil is heated under pressure, often with a catalyst. The large molecules of heavy oil crack, or break down, into the smaller ones found in petrol.

Cracking also produces a lot of light gases with even

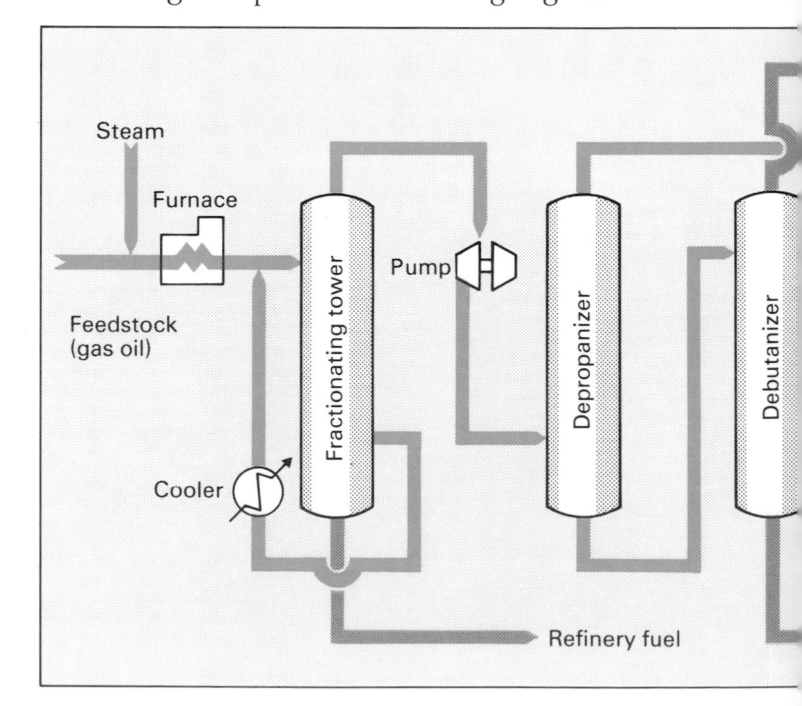

smaller molecules. In another important refinery process, polymerization (see page 62), these small molecules are combined into the kind of molecules found in petrol. Many other refinery processes change the liquids and gases obtained from petroleum into thousands of useful chemicals. In fact petroleum has become the biggest source of organic chemicals, often called petrochemicals.

A good example of how versatile petroleum chemicals can be is provided by a gas called ethylene, or ethene. It is a very reactive gas that combines readily with other chemicals. It can be converted into the plastics polyethylene and PVC, synthetic fibres, synthetic rubber, paint resins, alcohol, acetic acid, and ethylene glycol (antifreeze). A similar range of useful products can be obtained from the other main refinery gases propylene (propene) and butylene (butene).

Computer Control

Oil refineries process millions of tonnes of crude oil every year. Yet if you visit one, you will find few people there. The refineries seem to operate by themselves. Liquids slosh through pipes, steam hisses through valves, motors hum and instruments flicker. Everything appears to run smoothly, without any human help. This is because processes are largely under automatic control. A refinery is an excellent example of process automation (see page 65).

The "nerve centre" of the refinery is the control room, where a few people can be found. But it is not they who are in day-to-day control, but a computer. The computer issues instructions to heaters, pumps, valves and motors throughout the plant to keep things running smoothly. And it can take instant action if something goes wrong.

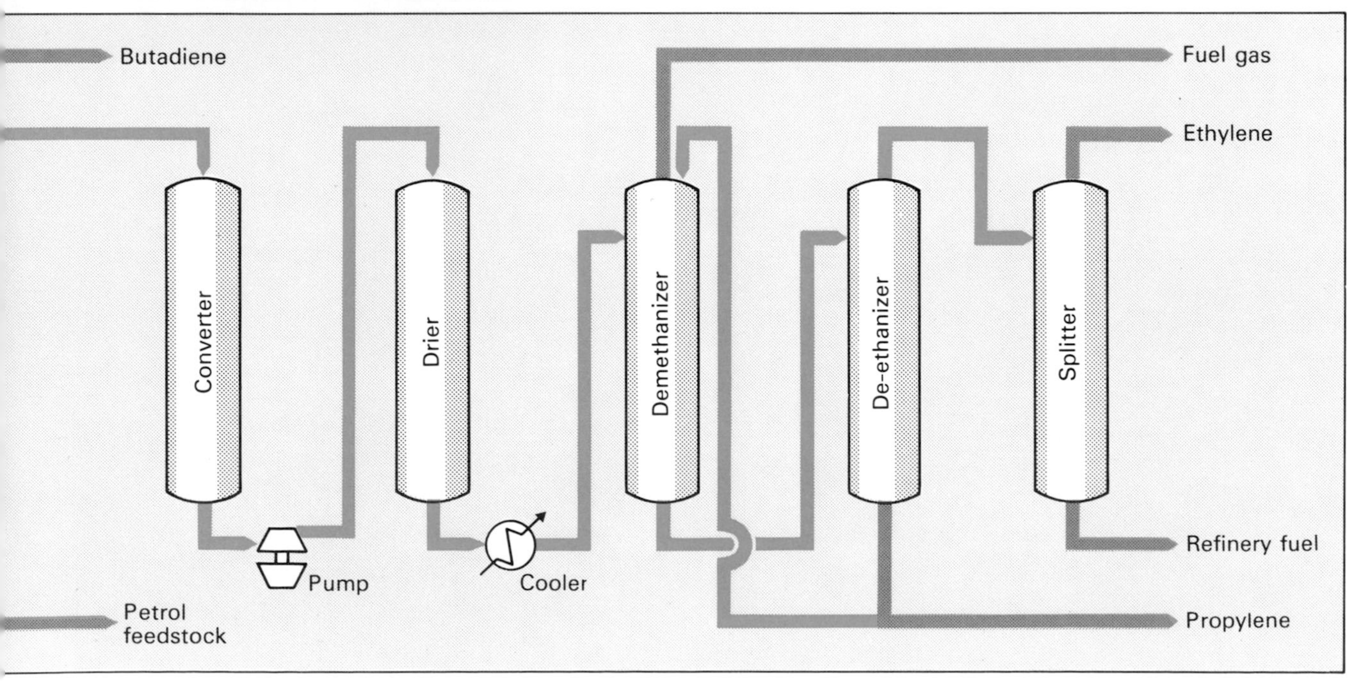

Chemical Industry 1

Most of the materials and substances we use in our everyday lives are not found in nature. They are man-made. They may be made, like rayon (page 52), by processing natural materials, such as wood, with chemicals. Or they may be made wholly from chemicals; such products are termed synthetic. Plastics like polythene and PVC are familiar synthetic products.

The chemicals required for the processing and synthesis of the products we use are supplied by a vast chemical industry. Major products of the chemical industry include heavy chemicals, plastics and synthetic fibres (see page 53), dyes, soap and detergents, explosives, pesticides and drugs.

The term "heavy chemicals" is used to describe chemicals produced in quantities of tens of thousands of tonnes a year. Most important is sulphuric acid, of which over 80 million tonnes are produced each year. Other heavy chemicals include sodium carbonate (soda), sodium hydroxide (caustic soda), ammonia and benzene.

There is a basic difference between the chemicals benzene and soda. Benzene is an organic chemical, and soda is inorganic. Organic chemicals were given their

Above: The control room in a chemical plant. Sensors installed throughout the plant measure such things as temperature, pressure and fluid flow. Their measurements are then displayed on gauges in the control room. The controller can then see at a glance what is happening anywhere in the plant.

Left: Tall towers dominate most organic chemical plants. It is in such towers that vital processes like distillation, cracking and polymerization take place.

Below: The chemistry behind industrial processes is worked out first in research laboratories like this one at Atsugi in Japan.

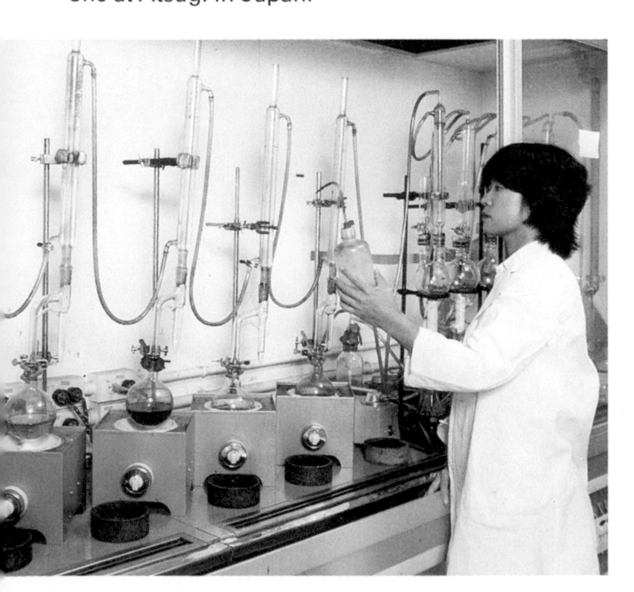

name because it was once thought that they could be produced only by living things. But we can now produce them in other ways. And we use the term "organic" to refer to compounds of carbon. Benzene is a compound of carbon and hydrogen. Most organic chemicals today are obtained from oil refineries (see page 56). COAL TAR is another rich source of organic chemicals that will be increasingly exploited when supplies of oil dry up.

We use the term "inorganic" to refer to compounds containing the other chemical elements. So caustic soda, which is a compound of sodium, oxygen and hydrogen, is an inorganic chemical. We obtain our inorganic chemicals from minerals in the Earth's crust and the oceans, and also from the air around us.

Chemical Reactions

The chemical industry makes use of all kinds of chemical reactions to make its products. Polyethylene plastic is made by a reaction involving only one material – a gas called ethylene. Ammonia is made by a reaction between nitrogen and hydrogen. Sulphuric acid is made by a series of reactions. Sulphur is first oxidized (combined with oxygen) to form the gas sulphur dioxide. This gas is further oxidized into sulphur trioxide gas, which is then, in another reaction, combined with water.

But in order to make ethylene change into ordinary polyethylene, it must be heated under very great pressure. The temperature needed is quite modest (about 200°C), but the pressure needs to be 1000–2000 times the pressure of air in the atmosphere! Lower pressures can be used if certain materials called CATALYSTS are present. Catalysts are substances that help chemical reactions take place. But they do not change chemically themselves.

A great many more chemical reactions cannot be carried out satisfactorily without heat, pressure and a catalyst.

Chemical Engineering

A chemist may be able to carry out a chemical reaction in a laboratory quite easily, using simple apparatus. But there is a big difference between a laboratory and a full-size chemical plant. For example, in the laboratory you can heat up a few drops of liquid in a test-tube, using a Bunsen burner. In industry, you have to find ways of heating up millions of litres of liquid.

The people who have to solve such problems are the chemical engineers. They take a process that works in a laboratory and try to make it work on a large scale. They design suitable reaction vessels, pumps, heaters, pipes, valves and so on, for handling and processing the raw materials and products. Usually, they first build a small-scale, or pilot plant. If this works satisfactorily, they then go ahead with the full-size one.

Chemical Industry 2

Dyes

Dyes have been used for colouring textiles for at least 5000 years. Early dyers used natural dyes obtained from the roots and leaves of plants, such as indigo (blue) and madder (red). Today, dyers use synthetic dyes, made from organic chemicals. W. H. Perkin made the first synthetic modern dye in 1856, when he extracted a mauve dye from coal tar. His work prompted other chemists to experiment with coal tar and the organic chemical industry was launched.

Mauve is one of a large range of dyes called aniline dyes. Another group are the azo dyes. These and other synthetic dyes cover nearly all shades of the rainbow. They are more vivid than the natural dyes. They are also more light-fast – they resist fading and washing out.

Explosives

Gunpowder, made from sulphur, charcoal and saltpetre (potassium nitrate), was invented by the Chinese more than 1000 years ago. It remained the only explosive until the mid-1800s, when nitroglycerin was discovered. But it only became safe to handle in 1867, when Alfred Nobel combined it with a kind of clay and produced dynamite. (In his will Nobel set up a foundation to fund the Nobel prizes awarded every year.)

Dynamite, largely used in mining and tunnelling, needs a shock from a detonator to set it off. It is very much more powerful than gunpowder and is called a high explosive. Trinitrotoluene (TNT) is another powerful high explosive, widely used in conventional (non-nuclear) bombs. The high explosive used in plastic explosives is made from ammonia and is known as RBX or cyclonite.

Pesticides

One of the reasons why farming is so efficient these days is because pests, such as insects, fungus diseases and weeds, can be controlled. Farmers have at their disposal a wide variety of chemical insecticides, fungicides and herbicides (weedkillers). These products are also available to the home gardener.

Chemists make many of their pesticides from organic chemicals containing chlorine or phosphorus. Among the chlorine-based compounds, usually called chlorinated hydrocarbons, are insecticides such as DDT, dieldrin, BHC and lindane; fungicides such as dichloran and benlate; and herbicides such as 2,4-D. 2,4-D is an example of a selective herbicide – it can kill broad-leaved weeds without harming grass, for example. The great drawback with most chlorine compounds is that they are poisonous to higher animals, including humans, as well as to insects. And they are very persistent – they remain in the environment for a long time after they have been applied,

Above: Fertilizer being added to the hopper of a seed drill. The chemical industry produces vast amounts of fertilizers, particularly superphosphate, made by treating phosphate rock with sulphuric acid.

Opposite top: Filling ampoules with a powerful pain-killing drug. In the pharmaceutical or drug industry, elaborate hygienic precautions are taken to prevent contamination of the products.

Opposite bottom: Standing on the ceiling is easy when your boots are stuck to it with modern epoxy-resin adhesives.

Below: Used in carefully controlled amounts, chemicals can successfully control the pests that attack crops without harming wildlife. Spraying from the air is an efficient way of treating large areas of land.

causing serious pollution (see page 68). For these reasons many especially deadly compounds, such as DDT and dieldrin, have been banned in Britain, the United States and several other countries.

Organic phosphorus compounds, called organophosphates, are now preferred to the chlorinated hydrocarbons because they are effective in much smaller doses, and they quickly lose their toxicity. This means that crops can safely be picked within a few days of application, and there is no poisonous build-up in the environment. Widely used organophosphates include malathion and parathion.

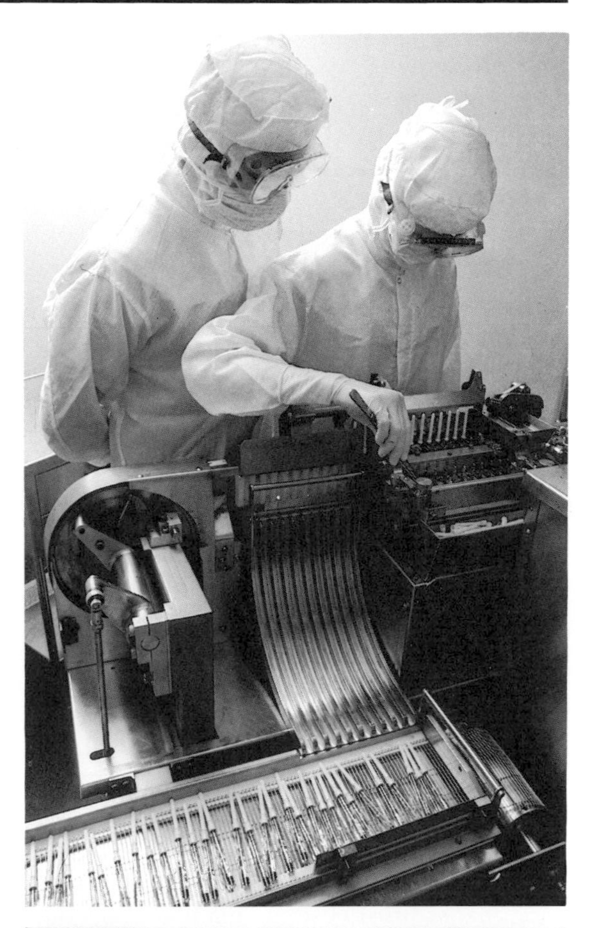

Drugs

These are the chemicals which some people take to relieve pain or treat ailments and diseases of the body. Through the ages people have known of the medicinal properties of certain herbs. But only in this century has the treatment of disease by chemicals – chemotherapy – become possible on a wide scale.

Traditional medicines like morphine and codeine, obtained from the opium poppy, are still widely used to relieve pain. Most drugs, however, are now made synthetically. The first and still the most widely used synthetic drug is aspirin, made originally from coal tar.

Chemists have produced a bewildering variety of other drugs to treat most of the illnesses that afflict the human race. They include anaesthetics (such as lignocaine) to kill pain; ANTIBIOTICS (such as penicillin) and sulpha drugs to kill germs; sedatives (such as barbiturates) and tranquillizers (such as Librium) to calm the nerves; and hormones (such as cortisone and insulin) to treat hormone problems. In recent years new techniques such as GENETIC ENGINEERING have been introduced to produce drugs.

Soaps and Detergents

These are substances that aid cleansing. Soap has been made for thousands of years by treating animal fats and oils with an alkali. Among the fats and oils most widely used today are tallow, coconut oil, palm oil, olive oil and soya-bean oil. The alkali used in most ordinary soaps is caustic soda. In soapmaking a mixture of fats and oils is boiled with caustic soda, producing crude soap and glycerin. The glycerin is an important by-product, being used to make antifreeze, plastics and nitroglycerin.

Soap, however, has the disadvantage of forming scum in hard water. Modern detergents form no scum. They have a better cleansing action than soaps. They increase the ability of the water to wet fabric fibres and "lift" grease more easily. They are made by treating compounds of benzene with sulphuric acid. Other chemicals may be added: fluorescing agents make clothes look brighter; ENZYMES in biological powders help shift stubborn stains.

Plastics

Left: The production of polyethylene sheet on an extrusion machine. The molten plastic is forced through a ring-shaped slot. On emerging, the plastic film is cooled and forms a large tubular balloon, which is eventually slit to form a sheet.

Right: Plastics abound in the modern car. The tyres are made from synthetic rubber. And in this MG Roadster the bumpers are as well. The steel body is plastic coated, for paints are now made primarily from synthetic resins. The sheen on the paintwork is achieved by the use of silicone polish; silicones are another type of plastic. The folding hood, once made of leather, is now made from polyvinyl chloride (PVC). Brushed nylon fabric covers the seats, which are upholstered with polyurethane foam. The carpets are woven from acrylic fibres.

Bottom right: The launching of one of the latest lock-out submersibles, widely used now in offshore-oil operations. Its hull is made from glass-reinforced plastic (GRP), popularly called fibreglass. The plastic is a kind of polyester.

Along with iron, concrete and wood, plastics have become among the most useful materials of our age. Over 50 million tonnes of plastics are produced every year, mostly from petroleum chemicals.

Plastics were first widely produced after World War II. At that time they were considered cheap and inferior substitutes for natural materials. But this is certainly not true today. Sometimes they are the only suitable material for a particular use.

Plastics can be found practically everywhere: inside and outside the home, for guttering, foam insulation, heatproof surfaces, easy-clean floors, "squeeze" bottles, bowls and cling film. In the form of synthetic fibres, plastics are found in the carpets, curtains and our "drip-dry" and crease-resistant clothes. Our furniture may be upholstered in plastic foam, and car tyres are made from plastics in the form of SYNTHETIC RUBBER.

Below: An illustration of the polymerization process. Under great heat and pressure, simple molecules of ethylene link together to form the polymer polyethylene.

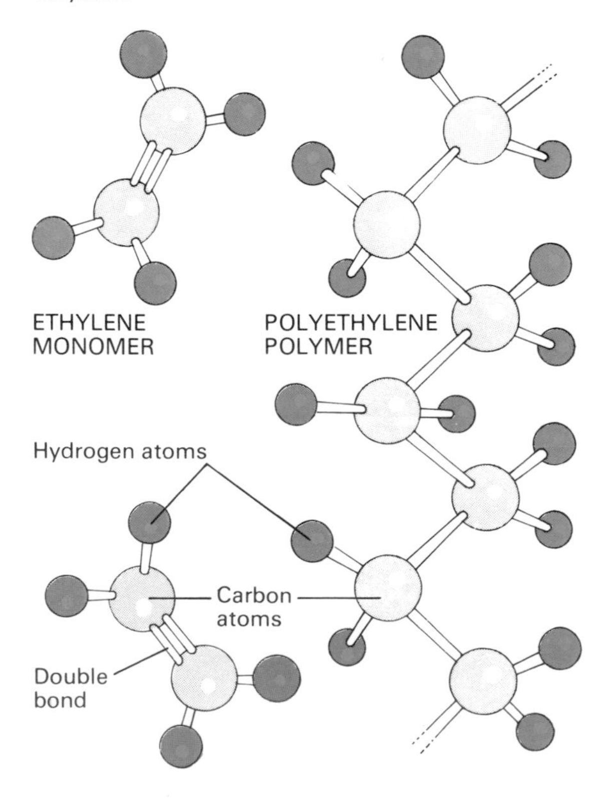

Polymers

But what exactly is a plastic? It is a substance which can easily be shaped by heating and which is made up of very long molecules. Most plastics have to have their molecules built up into chains from substances with smaller molecules. The long molecules of polyethylene (polythene) are built up from the simple molecules of ethylene, which contain only two carbon atoms. The chemical process in which such a build-up occurs is known as polymerization. The product, such as polyethylene, is called a polymer.

"Polymer" means "many parts". The name of most plastics begins with "poly", followed by the name of the substance from which it is made – for example, polyethylene and polyvinylchloride (PVC). Some plastics are better known by a common name, such as NYLON (a polyamide), or a trade name, such as Dacron (a polyester).

Thermoplastics and Thermosets

The plastics mentioned so far become soft when they are heated. They are known as thermoplastics. Other plastics may be shaped by heating, but then set rigid. They are known as thermosets, or thermosetting plastics.

Thermoplastics are in general quite soft and flexible. They are ideal for making such things as squeeze bottles and plastic bags. They are also suitable for making into textile fibres. Synthetic fibres made from nylon, polyesters and acrylic plastics are very widely used.

The best-known thermosetting plastic is Bakelite. It is important historically because it was the first synthetic plastic made, in 1909, by Leo Baekeland in the United States. Bakelite is made from phenol and formaldehyde, which link together to form a polymer. The atoms not only link to form long molecules, but link with atoms in other molecules. This cross-linking keeps the molecules firmly in place and makes Bakelite a strong, hard and rigid material, even when heated. Bakelite is not only heat-resistant, it is also an excellent electrical insulator.

Thermosets are made in two stages. First, they are produced in the form of a resin-like material and are known as synthetic resins. Then they are heated and set. Many other kinds of synthetic resins are made and used for a variety of purposes.

Polyurethane resins are used in paints and varnishes that give a tough and scratch-resistant finish. Epoxy resins are used to make powerful adhesives. They are used, for example, to bond metal in planes. The new "superglues" are acrylic resins that set instantly when applied. Polyester resins are used with fibreglass to make plastic car bodies and boat hulls.

Shaping Plastics

Plastics can be shaped in a variety of ways. Most are made by moulding. Some are blow-moulded – molten plastic is blown into shape in a mould. Some are injection moulded – molten plastic is injected into a mould. Thermosets are compression moulded – resin is pressed into shape while it is being heated.

Plastic pipes are made by extrusion – molten plastic is forced through a hole in a die. Film is made by extruding the plastic through a fine slit. Some plastic objects are shaped by laminating – they are built up layer by layer. Others are cast into shape.

Mass Production

It costs a great deal of money to set up a factory or a chemical plant. For example, buildings have to be bought or rented; machinery has to be purchased; and a labour force must be employed to run it. The money manufacturers make by selling the goods must pay for the costs of running the factory – for wages, raw materials and power. It must also help to pay back the money laid out, or invested, in setting up the factory in the first place, which probably came from a bank.

Only by producing certain goods, such as cars, in large quantities can manufacturers pay for all these costs, make a profit, and still keep prices down to a level that people can afford. The process of manufacturing products in large quantities at low cost per product is known as mass production.

Above: Part of the production line of the Triumph Acclaim saloon, a joint British–Japanese design. Body shells are being transferred by overhead conveyer for further welding operations.
Below: An automated laser cutting machine. It works under computer guidance, controlled by coded numerical signals on magnetic tape. This is termed numerical control.

Specialization and Mechanization

A manufacturer achieves mass-production by two main means. One method is to divide up the job to be done into a series of relatively simple tasks. Each worker is given one of these tasks to carry out. Because it is simple, he can do it very quickly. This way of organizing work is called the specialization of labour.

The second method is by using machines. This is called mechanization. By using machines, workers can work much faster than they could using only their hands.

Revolution in Industry

Until about 200 years ago workers had few machines to help them. Then in the 1760s a number of machines came into use to speed up spinning. The spinning machines were installed in factories and workers were employed to operate them. Within a few years, spinning had changed from a home craft into an industry.

It was the beginning of the period of history we call the Industrial Revolution, or the Age of Machines. Machines gradually came into use in other industries and they were powered by that most important machine, the steam engine.

Precision Tools

The successful development of the steam engine depended as much as anything on the ability of engineers to produce accurate, well-fitting parts. The pistons, for example, had to fit tightly in the cylinders to prevent leaks and loss of power. Well-fitting parts could now be produced, thanks to the invention of accurate boring machines and lathes. These were the forerunners of today's precision machine tools (see page 51).

A precision machine tool can turn out a virtually identical part each time, accurate to tiny fractions of a millimetre. All the parts it produces are interchangeable with one another. This makes possible mass production on

an assembly line. In this method of production workers (and sometimes robots) stand in line and add different parts to a product as it passes slowly past them on a CONVEYOR. This type of operation is used in car manufacture, still one of the best examples of mass production.

Automation

Many machines still need a human worker to operate them, for example to guide the cutting tools on a lathe. But more and more machines are being introduced that work automatically. All the workers need do is switch on the power. The machine will then operate automatically according to instructions fed to it, usually by a computer. The computer not only tells the machine what to do, it also checks that the machine does what it is told. If the machine starts to err, the computer can correct it immediately or stop the process until the fault is rectified. In this way a high level of accuracy is maintained.

The use of such automatic, computer-controlled machines is popularly known as automation. It is the latest stage of the Industrial Revolution. Automation can perhaps best be seen in chemical plants, such as an oil refinery (see page 56). Other industries are more difficult to automate. But, thanks to the invention of the silicon chip, computers are becoming so small that they can be installed in virtually any machine. These machines then become robots that can to some extent think for themselves (see page 66).

Below: An assembly "cell" in a robot-making factory at Yamanashi, near Tokyo in Japan. In this factory robots assemble other robots. If you look carefully, you can see this happening in the picture.

Robots

The latest development in factory production has been the introduction of robots. In several industries human workers and robots now work side by side on the factory floor. In the car industry, for example, robots weld car bodies and spray them with paint.

In the Western world the American company Unimation pioneered industrial robots in the 1960s. But it is in Japan that robots have been most widely used. By 1983 Japan had over 15,000 advanced industrial robots, compared with only about 5000 in the United States and about 4000 in Western Europe. One Japanese company, Fujitsu Fanuc, even has robots making other robots! Japanese workers work happily alongside the robots and give them pet names.

Popularly, robots are thought of as mechanical human beings – mechanisms that we call androids. But it is difficult and usually unnecessary to make robots in human form. Industrial robots are devices built to carry out work that would otherwise be done by humans. Their shape depends on what they have to do.

Above: Robots are now widely used on car assembly lines to weld the body shells and also to spray them with paint, carrying out tasks that are unpleasant for human workers.

Robot Anatomy

Industrial robots may not look like humans. But they must be similar in some respects if they are to replace humans on the factory floor. They will usually have "arms", "hands" and "muscles", and also a "brain" and a "memory". They must also have a kind of "nervous system" so that messages can be sent to control their actions.

The "arms" and "hands" of a robot are made of metal. Its "muscles" are electric motors or hydraulic (liquid-pressure) systems. Its "brain" is an electronic brain, or computer, which has a memory built-in. Its "nervous system" is a network of copper wires and printed circuits, which carry messages in the form of electric currents.

Before such a robot can take its place on the production line, it must be taught what to do. Its teacher is a human worker, who guides the robot through the precise motions it has to perform. The robot memorizes what it has to do, and can then work independently.

Most of the present generation of robots work very accurately, but they are blind, and this limits their use. The next generation of robots will have electronic "eyes". They will be able to recognize objects, pick them up and position them properly. This will make them suitable for many more assembly tasks.

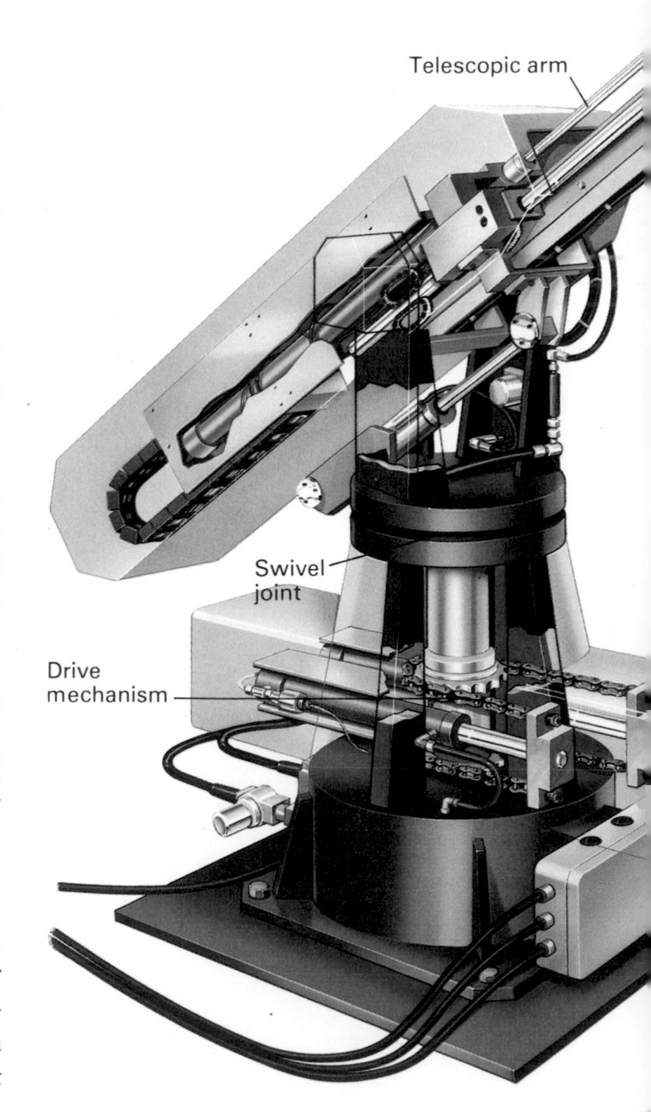

Tireless Workers

There are many advantages of using robots in industry. For one thing, they never tire. They can work non-stop for day after day, with only an occasional break for maintenance. And their work is always of the same high standard. When human workers work long hours, they get

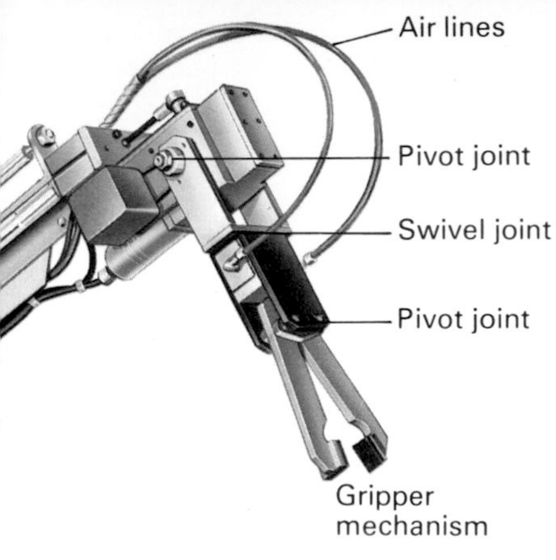

tired and their standard of work falls. Also, robots do not get bored repeating the same action all the time. Human workers soon become bored when they have a repetitive, monotonous job to do.

Another great advantage of using robots is that they can work in conditions that humans would find unpleasant. Robots do not mind noise, glare, changing temperatures, choking fumes and dangerous radiation.

Although robots are expensive to buy, they soon prove economical because they can work round the clock. And as more robots are made, their cost will fall. Human labour, on the other hand, is getting more and more expensive. So in the long run the greater use of robots should bring about greater productivity, lower costs, and also free human workers from boring, monotonous work.

Telechirics

Many semi-robot devices have been in use in industry for many years. In the nuclear industry workers use remote-controlled manipulators to handle dangerous radioactive substances. They control the manipulator's "arms" and "hands" with their own hands by means of a special hand grip. Such devices are often called telechiric ("hands at a distance") devices.

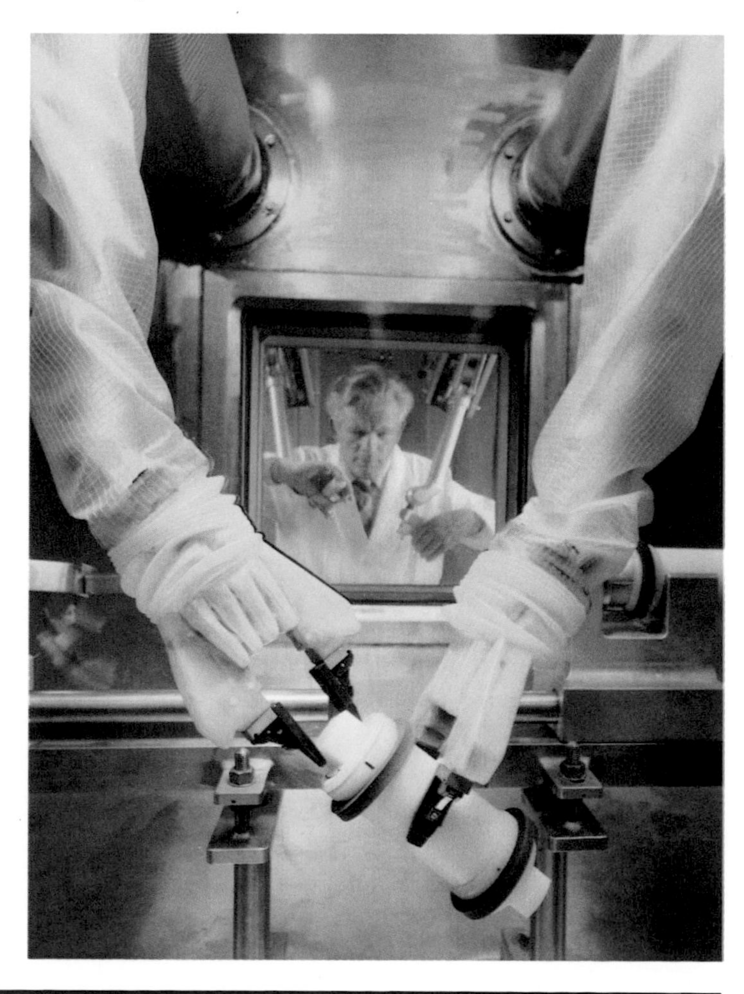

Top: The telechiric miner, a remote-controlled robot device that could be used one day to dig coal from seams too difficult or too dangerous to be mined by humans.

Left: A cutaway of a one-armed industrial robot, which looks rather like an ungainly bird. It can swivel its "body" round, move its "arms" up and down and in and out, twist its "waist" and grip with its "fingers". In this way it can duplicate many actions of the human frame.

Right: A nuclear-power worker operates a remote manipulator arm to handle dangerous radioactive material. Robot devices are ideally suited to such hazardous operations.

Pollution

We now live in a consumer society. We are able to buy and enjoy a large variety of goods, from canned foods to motor cars, and much of what we buy is later thrown away.

This is not only very wasteful of resources, it also poses a problem; what to do with all the rubbish. And a colossal amount of rubbish is involved. In Europe on average each person gets rid of over one kilogram of rubbish a day. In the United States, it is nearly three kilograms for each person. Unfortunately, some of the waste gets thrown away on "junk" heaps both in the towns and in the countryside. Litter spoils our surroundings – our environment. It is a form of pollution.

The factories that make the goods we use also produce wastes. Waste gases from factory chimneys escape into the air. Liquid wastes may escape into the rivers. These wastes again spoil our environment – they cause pollution.

There are many other sources of pollution in our civilization including pesticides (see page 60) and radioactive wastes (see page 18). But perhaps the most serious sources at present are acid rain, car-exhaust fumes and oil pollution.

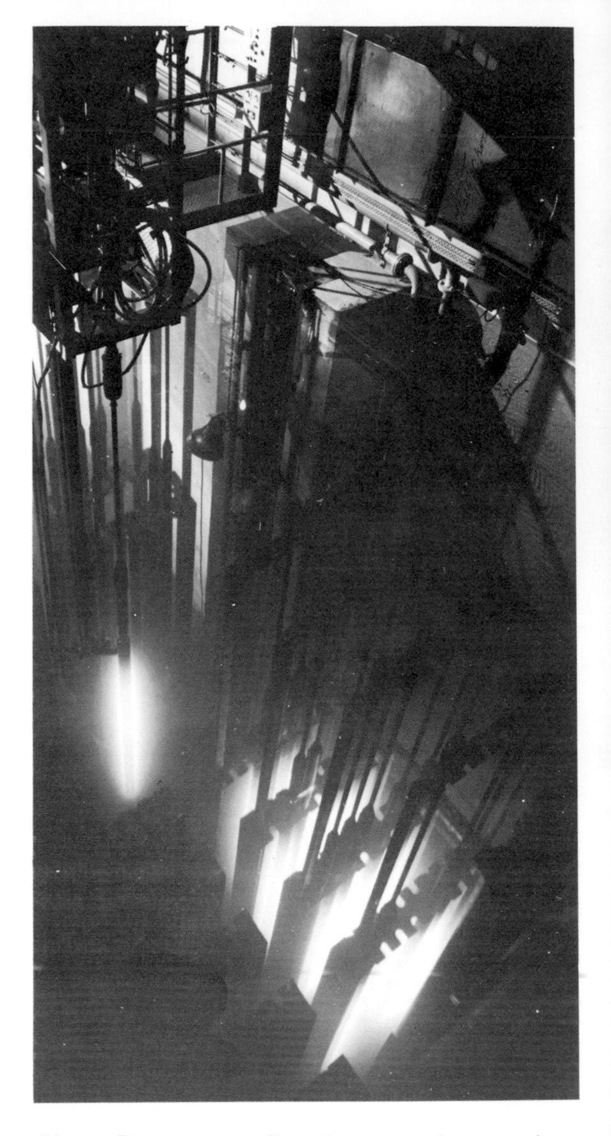

Above: Dangerous radioactive waste from nuclear reactors is kept in deep storage ponds for some time until its radioactivity is greatly reduced. The radiation it gives off creates the bright glow you can see in the picture. In the air, the radiation is invisible, so there is no warning of its deadly effect. The nuclear power industry thus takes elaborate precautions when handling and disposing of radioactive waste to prevent radiation escaping into the environment.

Acid Rain

Factory chimneys pour forth not only smoke, but also gases such as sulphur dioxide and nitrogen oxides. The smoke tends to fall out of the air locally, making industrial areas rather mucky places. The gases, however, can be carried much farther afield. And they pose a much more serious threat to the environment. This is because they combine with moisture in the atmosphere to form sulphuric and nitric acids. When it rains, the rain is acidic.

Many experts believe that acid rain is the worst pollution problem of all. It has already turned hundreds of lakes in the north-eastern United States, Canada and northern Europe into lifeless bodies of water, where fish and plants cannot survive. Thousands more are badly affected. In these areas there is no lime in the rocks to neutralize the acidity naturally.

In Sweden they pump lime into the acid lakes to try to reduce the damage. But this cannot be done on a large enough scale to save all the lakes. The answer lies in beating the pollution at its source – in the factories. This can be done, for example, by installing "scrubbers" which spray the chimney fumes with limewater.

Car-exhaust Fumes

A source of air pollution closer to home is the motor car, whose engine burns petrol. The fumes from the car exhaust contain specks of soot, carbon dioxide, carbon monoxide, and usually traces of lead. The carbon monoxide and lead are highly poisonous. In the middle of cities

Right: A guillemot covered in oil, a pitiful sight that is tragically becoming only too common along our coasts. When lightly oiled, birds can be washed in detergent and saved. But birds like this one are beyond help.

Above: A grim warning posted at the town of Seveso, near Milan in Italy, contaminated by the dangerous pesticide dioxin in July 1976. The population had to be evacuated from the town when a cloud of dioxin dust escaped from a nearby chemical factory.

car-exhaust fumes can build up to cause a health hazard to the people.

Pollution becomes much more serious in certain places where the air becomes trapped and stagnant. This happens in Los Angeles in the United States. There, smoke and exhaust fumes often get trapped in moist air and form a dense, choking fog known as smog. Smog can be a killer.

To combat the problem, governments are passing laws to reduce the amounts of poisons that can be emitted in a car exhaust. This may involve removing lead from the petrol, redesigning the engines, or fitting emission-control devices. A better answer in the long run will be to produce engines that use fuels like hydrogen which cause no pollution. Or perhaps only electric cars should be allowed in cities.

Oil Pollution

Most holidaymakers know only too well the effects of oil pollution on the beaches. Much of the oil comes from ships illegally cleaning out their fuel tanks. Sometimes offshore oil wells "blow out" and discharge vast amounts of oil into the sea. This has happened in the North Sea and in the Gulf of Mexico. Oil tankers may be wrecked, spilling their cargo into the sea.

One of the worst tanker disasters was the *Amoco Cadiz* (1978), which ran aground off the Britanny coast and spilled over 200,000 tonnes of oil. This polluted hundreds of kilometres of beautiful coastline. In particular it caused an appalling loss of life among seabirds, shellfish and other marine and coastal life. Perhaps this is part of the price we have to pay for modern technology.

Chapter 4

Construction

Construction engineers can be found practically everywhere, changing the face of the Earth. They blast tunnels through mountains to carry roads and railways. They build graceful suspension bridges to span rivers and estuaries. They raise great walls of concrete to dam the flow of mighty rivers. They erect towers that soar into the clouds. They also build runways, harbours, canals, factories and power stations.

Construction engineering is no delicate technology. Everything takes place on a massive scale. Millions of tonnes of soil and rock often need to be shifted before construction can begin. And millions of tonnes of concrete and steel are often required during construction.

Work continues day and night on many construction projects to ensure completion on time. Here construction engineers are finishing off a surge shaft for the Dinorwic pumped-storage scheme in Wales. The shaft measures 10 metres across and is sunk 439 metres into the mountain.

Muck Shifting

Man has been changing the face of the Earth since the beginning of civilization. Even 4000–5000 years ago he was building roads and bridges to improve communications, canals for irrigation, and raising impressive temples and tombs to the glory of his gods and dead kings. The construction of massive structures such as dams, bridges, skyscrapers, roads, tunnels and canals now forms part of the work of the civil engineer.

Civil engineering was originally so-called to distinguish it from military engineering. It became the first of the great engineering professions. The first national Institution of Civil Engineers was founded in England in 1828, and the celebrated road, bridge and canal builder Thomas Telford became its first president.

In the royal charter of the Institution occurs what is perhaps the best definition of civil engineering: "The art of directing the great sources of power in nature for the use and convenience of man."

In most kinds of construction work one of the biggest tasks is earth-moving, or as civil engineers call it, muck-shifting. Unbelievably large quantities of "muck" have to be shifted sometimes, particularly in the construction of dams and canals.

The New Cornelia Tailings Dam in Arizona, completed in 1973, was built from over 200 million cubic metres of soil and rock. In the Sudan, in Africa, a canal called the Jonglei is being excavated, which requires the removal of over 1000 million cubic metres of material. The Jonglei Canal is being built to divert the waters of the White Nile for irrigation. When completed, it will be over 350 km long – more than twice the length of the famous SUEZ CANAL.

Above: A self-propelled scraper removing the top-soil during the construction of a motorway. Extra power for the operation is provided by a bulldozer-tractor pushing at the rear.

Below: An excavator-loader working in conjunction with a dumper truck. The truck is ruggedly constructed for off-the-road operation, and has a simple tipper body for easy discharge of its load.

Muck Shifters

To shift these vast amounts of "muck", engineers use a variety of powerful earth-moving equipment. This includes excavators, power shovels, bulldozers and scrapers. The excavators include draglines, which are machines like cranes with a long arm or boom. They fling out a toothed bucket rather like a fisherman casting his line. The bucket is then dragged back over the ground and scoops up the soil. Some of the largest draglines are seen, not on construction sites, but at opencast mines (see page 42). For projects like the Jonglei Canal, bucket-wheel excavators are often used, which excavate with a huge rotating wheel with buckets around the edge. This type is also used in many opencast mines.

Other types of excavators, familiar on building sites, have digging buckets on hydraulic arms. They may also have an arm with a shovel attached. Some have caterpillar, or crawler tracks, to enable them to move more easily over rough and soggy ground. The bulldozer has crawler tracks

too. It is a powerful tractor with a blade at the front which is used for shifting soil and removing tree-stumps and other obstacles.

For shifting soil in bulk a scraper is used. This is basically a huge bowl with a cutting blade and an opening at the front lower edge. When it moves across the ground, the blade slices into the soil, which is then forced into the bowl. The big scrapers are self-propelled. They have powerful diesel engines, one at the front in a tractor unit, and one at the rear. A flexible joint between the tractor and bowl allows these units to swivel and roll independently and reduces the risk of the tractor overturning.

Firm Foundations

After excavation has been done, the next main task is usually building foundations. Every large structure – bridge, skyscraper or dam – requires firm support, or foundations. If there is solid rock near the surface, this is ideal. If not, then engineers have to provide a suitable alternative. The type of foundation they choose will depend on the properties of the soil, which are therefore investigated by experts in SOIL MECHANICS.

As a result of soil tests, engineers may decide on a raft foundation – a huge block of reinforced concrete. Or they may choose pile foundations. These consist of columns of steel or reinforced concrete which are driven (by pile-drivers) or cast deep in the ground. For bridge foundations under water, CAISSONS may be used. These are prefabricated structures that are sunk into the river bed and then filled with concrete.

Below: A crawler-tracked excavator of a type widely used on construction sites. It is powered by a diesel engine. Its digging arm is moved by hydraulic jacks, which are worked by liquid pressure. The motor that drives the crawler tracks also works hydraulically.

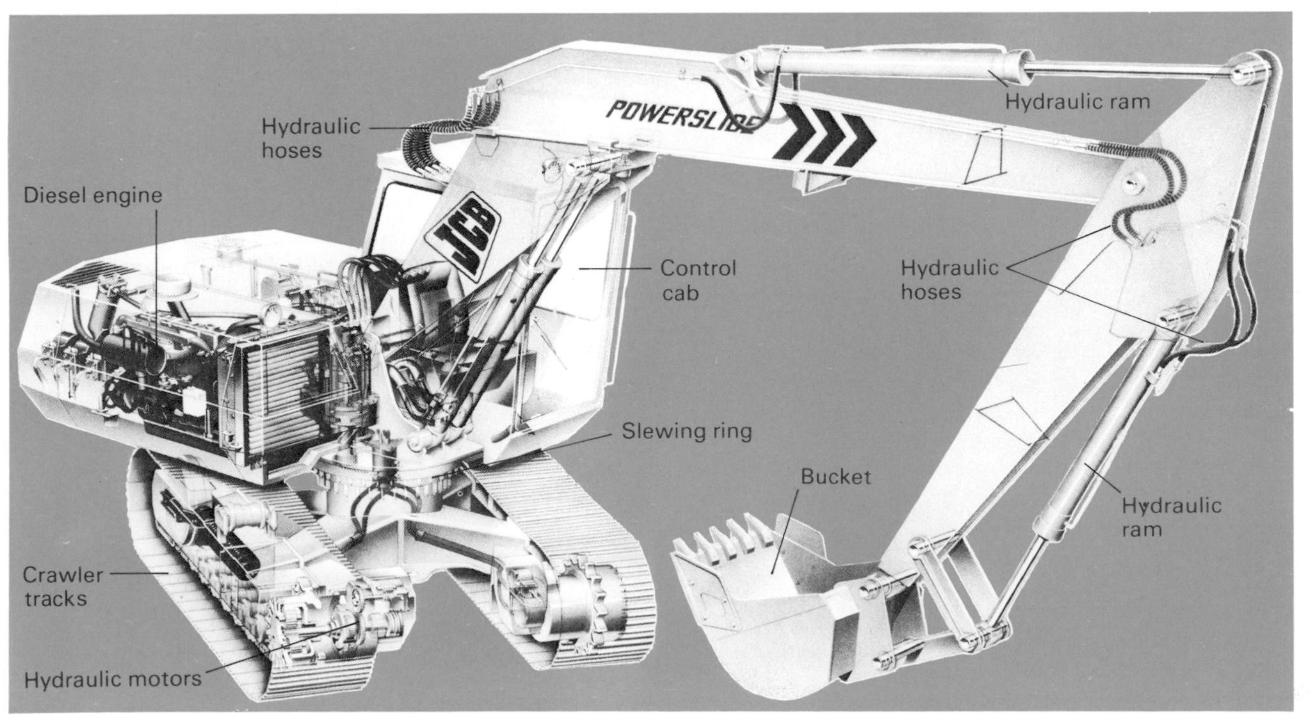

Skyscrapers

In most of the big city centres of the world, office blocks are soaring higher and higher. They are being built upwards rather than outwards because city land has become scarce and very expensive. Supreme among these aptly named "skyscrapers" is the Sears Tower in Chicago. It soars to a height of no less than 443 metres and has 110 storeys. In fact, its floor area is over 100 times the ground area on which it is built.

It is appropriate that Chicago has the world's tallest building for that is where the first modern skyscraper, the Home Insurance Building, was built in 1885. For most of the time since 1930 New York City has boasted the tallest skyscrapers, notably the 381-metre high Empire State Building. This is now the city's second tallest building, about 35 metres shorter than the twin towers of the World Trade Center.

Building high is not just a modern trait, however. As long ago as 2600 BC the Great Pyramid of Cheops was erected at Giza, in Egypt. It was built of over two million limestone blocks and reached a height of 146 metres from a perfectly square base with sides of 230 metres.

To the ordinary people of the day, living in single-storey mud huts, the Great Pyramid would indeed have appeared to scrape the sky. It still stands today as a monument to ancient construction technology.

Building Skyscrapers

One of the reasons New York is a skyscraper city is that solid rock can be found at or near the surface. So there are no problems about building foundations. In other cities engineers are not so fortunate. London, for example, is built on clay. To provide support for the 183-metre high National Westminster tower block, 375 piles had to be driven through the clay to stronger gravel beneath.

Most skyscrapers achieve their dizzy height thanks to the immense structural strength of steel. They are built with a frame of welded or riveted steel girders. This steel skeleton carries the weight of the whole building. The walls need only be thin and lightweight, for they take no load. In traditional building, the walls take all the load. This limits the height of such a building to a few storeys, otherwise the walls at the bottom would have to be impossibly thick.

Curtain Walling

The lightweight walls of a skyscraper act rather as curtains to shut out the elements, and form what is called curtain walling. They are often just made of glass in aluminium or stainless steel frames. To prevent the building acting like a greenhouse, the glass is often covered with a fine layer of gold (about 0·0005 mm thick), which filters out and reflects the Sun's rays.

The steel-frame/curtain-wall method of construction is also very fast. Once the frame has been erected, work can proceed on any level. It took only 15 months, for example, to build the 102-storey Empire State Building.

Above: The 189-metre high Post Office Tower in London. Radio and television relay towers like this are among the world's tallest structures.

Left: The Hancock Tower in Boston. Glass is a favoured material for curtain-walling.

Bottom left: These skyscraper hotels in downtown Los Angeles are walled with gold-tinted glass.

Below: The incomparable skyline of Manhattan, New York, dominated by skyscrapers. Tallest are the twin towers of the World Trade Center, which rise to over 415 metres.

Tower Cities

The Sears Tower has a working population by day of nearly 17,000, so it is like a city within a city. Some people have suggested that cities may one day be built in towers so as to house an expanding population, yet leaving plenty of space on the ground. The Dutch architect Wilem Frischmann reckons that a city of a quarter of a million people could live in a tower block 200 metres square and with 850 storeys. It would soar to a height of three kilometres! What is more, this kind of building is already possible with existing technology.

Roads and Bridges 1

The invention of the motor car about a century ago brought about a social revolution among a large proportion of the population. It affected where people lived and worked and what they did for relaxation. To cope with this new machine, the road systems of the world had to be greatly improved and expanded on a scale unknown since Roman times.

The Romans some 1900 years ago, were the first great road builders. They, like us, realized the advantages of good communications. From Rome they built networks of fine straight roads, totalling up to 80,000 km in length. These roads spanned the length and breadth of their vast empire, which stretched from Britain to Persia (now Iran), and from Germany to North Africa.

The Romans laid down many of the principles of road building still followed today. Their roads were made as straight as possible and consisted of layers of stones on hard-rammed soil. Often lime was added to one layer to form a kind of concrete. The road surface was curved upwards, or cambered, to allow the water to drain off, and there were usually drainage ditches at each side.

Superhighways

Not until the 1920s were such long, straight and carefully engineered roads built again. Appropriately enough, the first ones were built in Italy. They were the first of the modern special roads restricted to motor traffic and designed for high speeds. They are called various names in various countries – autostrada in Italy, autobahn in Germany, motorway in Britain, expressway or freeway in the United States.

Although they have different names, these superhighways have a similar design. They have several traffic lanes, and lanes going in opposite directions are separated. There are no traffic lights or roundabouts at the main road level. Other roads pass over the top or go underneath. Traffic enters and leaves the superhighway at ingeniously designed intersections, such as the common "cloverleaf".

As perhaps to be expected the United States, "home of the motor car", has the world's biggest road network totalling some 6 million km in length. It also has the biggest superhighway system, known as The Interstate, which runs for nearly 70,000 km. The longest highway in the system, Interstate 80, runs for 3000 km from New York City to Salt Lake City, in Utah. This is longer than all the motorways in Britain put together!

Road Building

In general the building of roads requires a great deal of ground to be levelled and soil to be shifted. This is where the great earth-moving machines – scrapers, graders, bulldozers and dump trucks – come into their

Left: Traffic speeds through the centre of Los Angeles on one of the city's many freeways, or motorways. This highway illustrates some of the basic principles of motorway construction. It has traffic lanes going in each direction, and opposing traffic streams are separated by sturdy crash barriers. There are no intersections with other roads at the same level.

Below: A concrete road being laid by a concrete paving train. In concrete road construction rotating-body mixer trucks pour ready-mixed concrete onto the road bed. Vibrating machines settle the wet concrete and then steel mesh is placed on top. Another layer of concrete is then poured over it and vibrated. Finally grooves are made in the surface to provide better grip for tyres.

own (see page 72). When the route has been levelled, it is made firm – ("compacted") by heavy "sheepsfoot" rollers.

On top of the firmly rolled ground a base course of gravel or crushed stone is laid, often mixed with dry cement. This is again rolled, and another course put down and rolled. The road is then ready to receive the top course, or pavement. This usually consists of a mixture of crushed stones and tar, known as tarmac or asphalt. But concrete pavements are also sometimes laid.

The tarmac is laid by mechanical pavers, which spread it into an even layer. Then a heavy diesel-powered roller comes along and rolls it. Concrete roads are laid by a so-called "concrete train". This consists of a series of machines that spread layers of concrete and vibrate them to make them firm. Steel-mesh reinforcement is added between the layers for strength.

Bridging that Gap

The major engineering work in a road-building project, however, may be the construction of bridges along the route. And bridge-building is one of the most challenging tasks of the civil engineer.

The Romans were also the first great bridge builders. Many of their bridges can still be seen today, often in ruins but still magnificent. Outstanding is the famous Pont du Gard aqueduct near Nîmes in the South of France, built some 1800 years ago. It is made up of three tiers of graceful arches.

These days a fascinating variety of bridges can be found carrying roads – and also railways – across river, valley and estuary. The commonest type is the simplest – the beam bridge. The bridge deck or roadway consists of a straight slab, or beam, supported at intervals by piers. This can span only small gaps. The arch design favoured by the Romans is also found widely and can span very much larger gaps.

Arch bridges are no longer built of stone as the Roman ones were, but of steel or reinforced concrete. Sydney in New South Wales, has two of the finest arch bridges. One is the slim and elegant Gladesville Bridge (1964), which has a concrete arch. The other is the huge Sydney Harbour Bridge (1932), which has a steel arch. The Sydney Harbour Bridge has a span of 503 metres, and a width of 48 metres, making it the world's widest. It carries two overhead railway tracks, eight traffic lanes, a cycle track and a footpath.

Steel-arch bridges like the Sydney Harbour are difficult and expensive to build. So are the great steel-truss bridges like the old Forth Rail Bridge (1890) in Scotland, which has twin main spans of 521 metres. Today there are better and more elegant ways of spanning such distances.

Above: A vertical-lift bridge across the Sacramento River in northern California. When a ship wants to pass, the bridge deck is raised vertically between the two towers.

Below: Rail and road bridges across the Firth of Forth in Scotland. The cantilever-type rail bridge (left) was completed in 1890, and has twin main spans of 521 metres. The suspension-type road bridge was completed in 1964, and has a 1006-metre span. It has the second longest span in Britain, after the Humber Bridge.

Cable Bridges

One of the most recent designs is the cable-stayed bridge. This consists of a straight bridge deck which is stiffened by means of cables attached to tall towers near the middle. Cable-stayed bridges were developed in Germany, one of the first being the Ebert Bridge over the Rhine near Bonn. Currently the largest cable-stayed bridge is the 404-metre span bridge over the Loire at St Nazaire in France.

For the longest spans, however, suspension bridges are chosen. In a suspension bridge the roadway hangs from thick cables, which go up and over tall towers built on either side of the gap being bridged. The suspension bridge with the largest span in the world (1410 metres), is the Humber Bridge in north-east England, opened in 1981. It is so long that the suspension towers are built slightly out of parallel, to allow for the curve of the Earth!

The next longest suspension bridges are in the United States. The Verrazano-Narrows Bridge (1298-metre span) is in New York; the Golden Gate Bridge (1280-metre span) is in San Francisco. The Golden Gate is the most famous and perhaps the most beautiful bridge of all.

The effect of the wind on such large bridges must be allowed for. In the past, bridges have collapsed because they could not withstand the extra loads and vibrations caused by strong winds. So these days bridge designers always test scale models of their designs in a WIND TUNNEL, just as aircraft designers do.

Above: One of the towers of the Golden Gate Bridge, which spans the entrance to San Francisco Bay, in California. The bridge was completed in 1937 and had the world's longest span until 1964, when it was eclipsed in size by New York's Verrazano-Narrows Bridge.

Top left: Switzerland has many spectacular bridges to match its spectacular scenery. This graceful arch rail bridge is the Langwieser Viaduct. The strength of the bridge lies in the arch shape. The sides of the arch transmit the load on the bridge down to the supporting foundations.

Building a Suspension Bridge

The main stages in construction are building the towers, "spinning" the cables, and assembling the roadway. The towers are made of steel or reinforced concrete and are built on massive piers of reinforced concrete. The Golden Gate Bridge has the tallest towers – 225 metres high.

The cables are made up of tens of thousands of strands of thin steel wire bound together. The laying, or "spinning" of the cables is done by means of a moving carriage. This is drawn back and forth up and over the towers and pays out strands of wire behind it. There are usually two cables on each side of the bridge. In the Verrazano-Narrows cables there is enough wire to stretch five times around the equator! At each end of the bridge the ends of the cables are fixed to massive anchorages of reinforced concrete buried deep in the ground. These have to withstand the enormous tension in the cables.

The bridge deck is suspended from the cables by thinner steel cables. It is assembled section by section either from the road level, or from underneath. The roadway of the Humber Bridge, for example, is constructed of "box girders" formed by welding together stiffened steel plates. It has a special AEROFOIL shape, rather like a plane's wing. This helps it resist the buffeting of the wind.

Dams

No other branch of construction, better fits the definition of civil engineering (see page 72) than dam-building. Dams do harness "the great sources of power in nature". They tame mighty rivers, often preventing large-scale flooding and loss of life. The water "stored" in the artificial lake, or reservoir, behind a dam may be used to provide hydroelectric power (see page 20). Or it may provide much needed irrigation for hitherto arid land. Some dams, for example the Tarbela Dam in Pakistan, are built to achieve all these objectives.

Dams are the biggest of all man-made structures. Built from concrete or just rock and soil, they can tower hundreds of metres into the air and stretch for many kilometres. The Grande Dixence Dam in Switzerland is the world's highest at 285 metres. The world's longest is the Kiev Dam in Russia, which stretches for over 50 km.

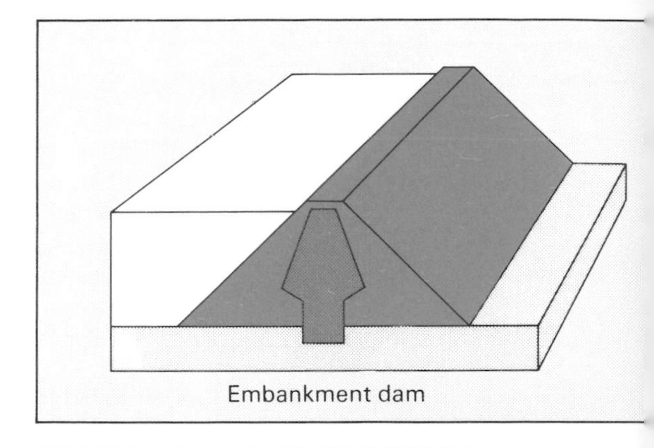

Embankment dam

The Itaipu Dam

Perhaps the most impressive feat of dam building is the Itaipu Dam in South America. It stretches for nearly 8 km across one of the world's longest rivers, the Parana, on the Brazil/Paraguay border. This dam consumed 11 million cubic metres of concrete – enough to build four or five new cities! At full power its hydroelectric plant will produce over 12 million watts of electricity – much more than any other in the world.

However, as in many dam projects, success has been achieved at considerable cost to the environment. The reservoir behind the dam has submerged some 1500 square kilometres of tropical forest and productive farmland. One of the world's finest waterfalls, the Sete Quedas, has disappeared. Some 40,000 people had to be moved from their homes.

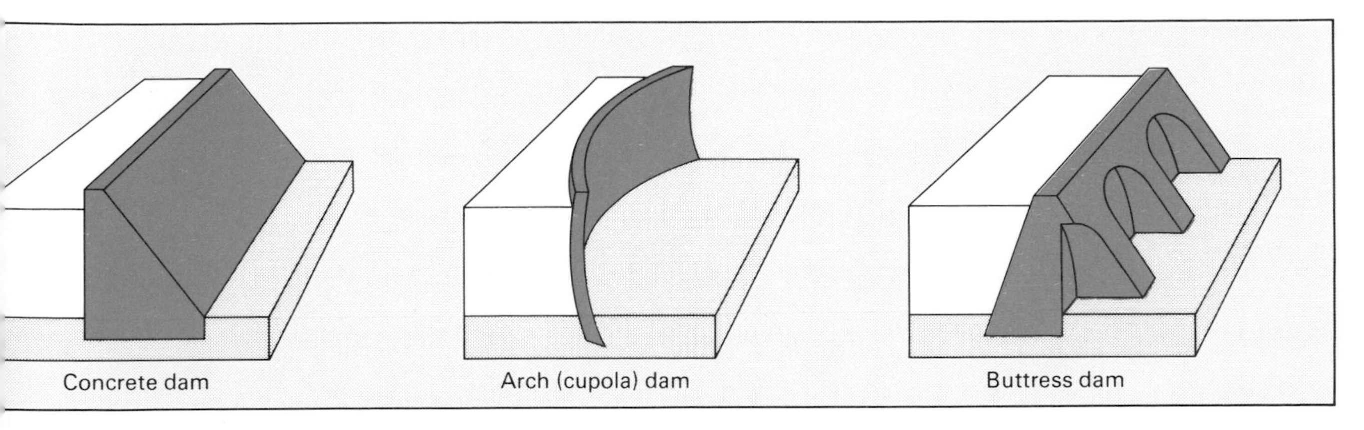

Above: Four types of dams in widespread use. The embankment and concrete gravity dams rely on their immense weight to keep them in place. The fragile-looking arch dam relies on its shape. The buttress dam is also economical on materials.

Left: The Mohammed Reza Shah Kabir Dam in Khuzestan, in south-west Iran. It is a cupola type, curving from side to side and from top to bottom.

Bottom left: A novel type of dam, the Thames Barrier. Built across the River Thames at Woolwich, it is designed to protect London from exceptionally high tides. The flood gates take the form of parts of cylinders that are rotated 90° into position when danger threatens.

Below: The Grande Dixence Dam in Switzerland, the world's highest dam since 1961. It is a gravity dam, 700 metres long and 285 metres high, constructed of 6 million cubic metres of concrete.

Gravity Dams

The Tarbela Dam was constructed from hundreds of millions of cubic metres of soil and rock. It is a type known as an earth-fill, or embankment dam, and is built by dumping load after load of rock and soil on top of one another until the desired height is reached. When finished the dam is very broad at the base, where it has to resist the greatest water pressure. It then slopes up on both sides to a relatively narrow crest. It is built with a central core of clay, or occasionally concrete, which extends deep into the ground. This design prevents water seeping through and underneath the dam.

The earth-fill dam is one type of gravity dam because it stays in place because of its enormous weight. Gravity dams can also be made of concrete. The gigantic Grand Coulee Dam in Washington State in North America is a concrete gravity dam. It has the cross-section of a right-angled triangle, with the perpendicular side facing the water. Again, its great weight (over 20 million tonnes) keeps it in place. In other concrete-dam designs a straight dam wall is strengthened by means of columns, or buttresses behind it.

Arch Dams

Concrete gravity dams consume an enormous amount of concrete at enormous cost. But sometimes the same effect can be achieved more economically by a change in design – by making the dam curve in the shape of an arch. Arch designs are particularly suitable for damming narrow gorges. The curve of an arch dam faces upstream, and the arch transmits the pressure of the water along to the sides. The sides must therefore be particularly well founded on rock.

A variation of the arch design is known as the cupola. The dam curves not only from side to side but also from top to bottom. Both the arch and cupola designs can be built very much slimmer than gravity designs. Not only do they use their shape to withstand the water pressure, they also exploit the most outstanding property of concrete – its strength under compression.

Tunnels and Pipelines 1

From prehistoric times people have burrowed into the ground like moles in search of riches of one kind or another. At first they tunnelled for flint, then later for other minerals: copper, gold, coal and diamonds. Those great engineers, the Romans, were the first to tunnel for other reasons when they hacked their way through solid rock to build aqueducts to carry water to their cities.

In the 1700s the first transport tunnels were driven, as networks of canals were built during the Industrial Revolution. Late in the following century the development of the railways led to the building of the first great modern tunnels, such as the Mont Cenis through the Alps.

Hundreds of kilometres of tunnels are built every year throughout the world. Many are for carrying water in hydroelectric power schemes, like that at Dinorwic in North Wales. There, a rabbit-warren of tunnels links the upper and lower reservoirs with the main machine hall, which is located in a mammoth cavern excavated in the rock. The cavern is twice as long and half as wide as a football pitch and higher than a 16-storey building.

Great tunnels are still being driven through the Alps. The latest is the 16-km long St Gotthard Tunnel from Göschenen to Airolo in Switzerland. Opened in 1980, it is the longest road tunnel in the world. At one point it is 2500 metres below surface level.

On the other side of the world, in Japan, great rail tunnels are being built. Currently the longest is the 22-km Oshimizu Tunnel on the island of Honshu. It carries a new extension of the Shinkansen – the high-speed railway on which the famous "bullet trains" run (see page 100).

Above: Inside the control cab of a tunnelling mole. Forward of the cab, hydraulic rams grip the sides of the tunnel while the cutting head advances.

Opposite: The Trans-Alaskan oil pipeline snakes through a beautiful autumn landscape. About half the length of the 1300-km pipeline is elevated on supports like this.

Bottom left: Laser beams are now being used to keep tunnelling machines on a straight path. The powerful beams are easily visible and remain perfectly straight.

Below: The entrance to the St Gotthard Tunnel through the Swiss Alps, since 1980 the world's longest road tunnel.

Tunnels and Pipelines 2

Modern Tunnels

Tunnelling has always been dangerous and remains so, even using modern technology. There is the danger of sudden rock falls, and the explosive inrush of water from flooded caverns or underground streams. Nineteen tunnellers were killed during the building of the St Gotthard Tunnel.

To minimize such dangers, geologists carefully survey a proposed tunnel route to check the properties of the rocks. Then tunnellers usually bore a small, pilot tunnel to make sure everything is satisfactory before driving the main tunnel. The method of tunnelling chosen will depend very much on the nature of the ground or rock through which the tunnel will pass.

Jumbos and Moles

The traditional method of tunnelling through hard rock uses explosives and can be described as "drill, blast and clear". First, holes are drilled into the rock face. They are filled with explosives, which are then set off, blasting the rock apart. The rock debris, or spoil, is then cleared away and the cycle starts again.

Today the drilling is done by hammer drills on a wheeled or tracked vehicle called a jumbo. The drills are mounted on hydraulic booms and are fed forward automatically as drilling proceeds. They can be controlled from a distance. After blasting, mechanical loaders load the blasted rock into dump trucks or rail cars for removal.

The cycle of drill, blast and clear takes a long time, and this has led engineers to develop continuous tunnelling machines. These are usually called moles. Some of the latest ones, called full facers, can bore tunnels up to 10 metres in diameter.

At the front of the mole is a broad cutting head, which carries a number of teeth or revolving discs. These grind away the rock when the head rotates. The rock spoil is carried from the face by a conveyor, which drops it eventually into waiting rail cars.

Behind the head, hydraulic rams grip the sides of the tunnel. This keeps the machine firm while the head is thrusting forwards. The front part of the mole also houses the control cabin, from which tunnellers "drive" the machine. It is surrounded by a steel cylinder, or shield, to protect the tunnellers against rock falls.

In soft and wet ground the tunnelling machine is supplied with compressed air. This helps prevent the tunnel caving in and keeps the water out. The latest machines use pressure only near the cutting head and use a liquid clay to remove excavated material. This avoids the need for the tunnellers in the control cabin to work under pressure all the time, which is very inconvenient and can be dangerous.

Above: A pair of drilling rigs used during the St Gotthard tunnelling project. Each rig carries four rock drills, mounted on hydraulic booms.

Below: Laying a section of the Trans-Alaskan Pipeline. The pipeline is constructed by welding together sections of 122-cm diameter pipe. Before burial, the pipe is coated first with tar and then with concrete.

The Japanese have gone even further and designed tunnelling machines which require no underground operators at all. They are controlled remotely from the surface, via closed-circuit television and use lasers to guide them.

Pipelines

Building pipelines is another task of the civil engineer. Pipelines are a very efficient method of transporting fluids over long distances. Many are used as aqueducts to carry water from reservoirs to hydroelectric power plants or cities. Others carry oil or natural gas between the oil and gas fields and refineries or ports, and from refineries to distribution centres. The United States has more than 300,000 km of oil pipelines alone.

Constructing pipelines is at first sight a much simpler process than tunnelling. Lengths of steel pipe are welded together and then laid in a trench, which is afterwards filled in. And where the lie of the land and the climate are favourable, it is as simple as that. But sometimes pipelines have to be laid under very difficult conditions indeed.

The most demanding pipeline projects are connected with the transport of oil and natural gas, which are often found in remote and inhospitable regions. To tap the oil in the North Sea, for example, many pipelines are laid on the seabed between the production wells and the land. Special ships have been built for the pipelaying, which is supervised underwater by divers and engineers in submersibles.

Underwater pipelaying was also the most difficult part of the Trans-Mediterranean Pipeline scheme, completed in 1981. This mammoth pipeline carries Algerian natural gas from Hassi R'Mel, in the Sahara Desert, over 2500 km to the Po Valley in northern Italy. The pipeline crosses the strait between Tunisia and Sicily, at depths of 600 metres.

Across Arctic Wastes

Problems of a very different kind were experienced on the building of the 1300-km long Trans-Alaskan Pipeline, completed in 1977. The pipeline snakes south across Alaska carrying crude oil from Prudhoe Bay to the ice-free port of Valdez. During construction, engineers had to battle against the climate as well as the difficult terrain, braving gale force winds, blizzards and temperatures down to −50°C.

Similar problems are encountered in Siberia, where work on the Trans-Siberian Pipeline is due for completion in the mid-1980s. The pipeline runs nearly 6000 km from Urengoi in northern Siberia and carries natural gas into the heart of Western Europe. At the same time five other pipelines are being laid to distribute Siberian gas throughout Russia, making a network of over 20,000 km of pipelines. Considered as a whole this becomes the largest construction project ever attempted.

Chapter 5

Transport by Land

Our main form of land transport, the motor car, has been with us for less than a century. During this time it has evolved from a slow, inefficient and bone-shaking horseless carriage into a swift, efficient and comfortable vehicle. Yet it still has its drawbacks, causing pollution of the air and, on a world scale, being the instrument of death for thousands of people every year.

Most goods are shifted overland by trucks, or lorries, which offer the convenience of door-to-door delivery. But for long-distance transport of goods in bulk nothing beats the train. Passenger traffic on the railways, however, has declined greatly over the years. But the introduction of a new generation of high-speed inter-city trains is gradually attracting passengers back.

To speed the flow of traffic on the roads, cleverly designed intersections have been built. This one is located at Gravelly Hill, Birmingham.

Petrol Engines

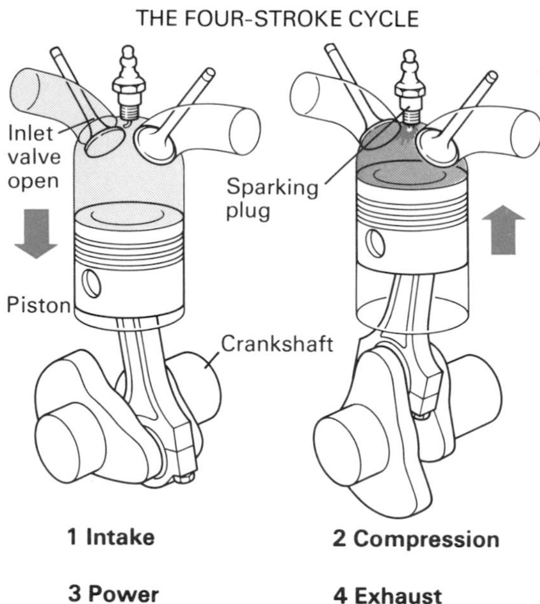

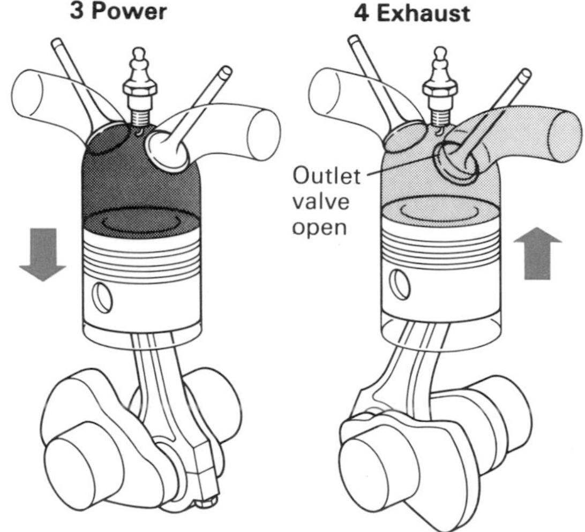

Above: The stages of the four-stroke petrol engine cycle. Intake: The piston falls and draws in fuel mixture. Compression: The piston rises and compresses the fuel. Near the top of the stroke, the fuel is ignited by a spark. Power: Expanding gases force the piston down, producing power. Exhaust: The piston rises again and pushes out the burnt gases. The cylinder is now ready to begin the cycle again.

Top right: An exploded view showing the main parts of a four-cylinder, in-line petrol engine. Most car engines are built in a similar way, the cylinder block and head being made of cast iron. The pistons drive round the crankshaft, spinning the flywheel. This connects with the car's transmission system.

Right: An entry in the Shell mileage marathon, in which ingeniously designed vehicles compete to find which can travel farthest on a certain quantity of petrol. Some vehicles have achieved the equivalent of travelling more than 650 km on a litre of petrol.

The most important source of power on the roads today is the petrol engine, which propels the majority of cars and motorcycles. It is not a particularly efficient engine, being able to use only about a quarter of the energy in the petrol to produce power. And it also produces poisonous fumes, which constitute a major source of pollution (see page 68). But it is likely to remain the prime source of power for cars for the time being.

Most petrol engines are reciprocating piston engines. They have pistons that move up and down in cylinders. One type of engine – the WANKEL – is a rotary engine. It produces power by means of a rotating rotor, but it is used at present only in a handful of cars.

Piston Power

In the ordinary piston petrol engine, petrol is mixed with air to form an explosive vapour and introduced into the engine cylinders. There it is exploded by an electric spark. The hot gases produced by the ignition of the fuel force down pistons and produce power. The pistons are connected by connecting-rods to a crankshaft and turn this round. At the end of the crankshaft is a heavy flywheel. This passes on the power to the car's transmission system, which drives the wheels.

In each cylinder certain actions are repeated according to a regular cycle so that the piston can continue to deliver power. It is called the four-stroke cycle, because the same actions are repeated every four strokes (upward and downward movements) of the piston. They are intake (or induction), compression, power and exhaust (see diagram left).

The most common type of petrol engine has four or six cylinders arranged in a row. This is called an in-line engine. Another popular arrangement is the V-engine. This has 4, 6, 8 or 12 cylinders arranged in two banks, inclined at an angle to each other in the shape of a V.

Engine Systems

The petrol engine is a complicated piece of machinery that includes as many as 150 moving parts. For easier understanding, we can think of the engine as being made up of a number of different systems – fuel, ignition, cooling and lubrication. Each plays a part in making the engine work.

The fuel system delivers the fuel to the engine cylinders. It pumps petrol from a fuel tank into a carburettor. There it is mixed with exactly the right amount of air to form explosive petrol vapour. This is then sucked into the engine cylinders. A more efficient method of getting fuel into the cylinders is petrol injection, in which a very precise amount of petrol is injected directly into each cylinder.

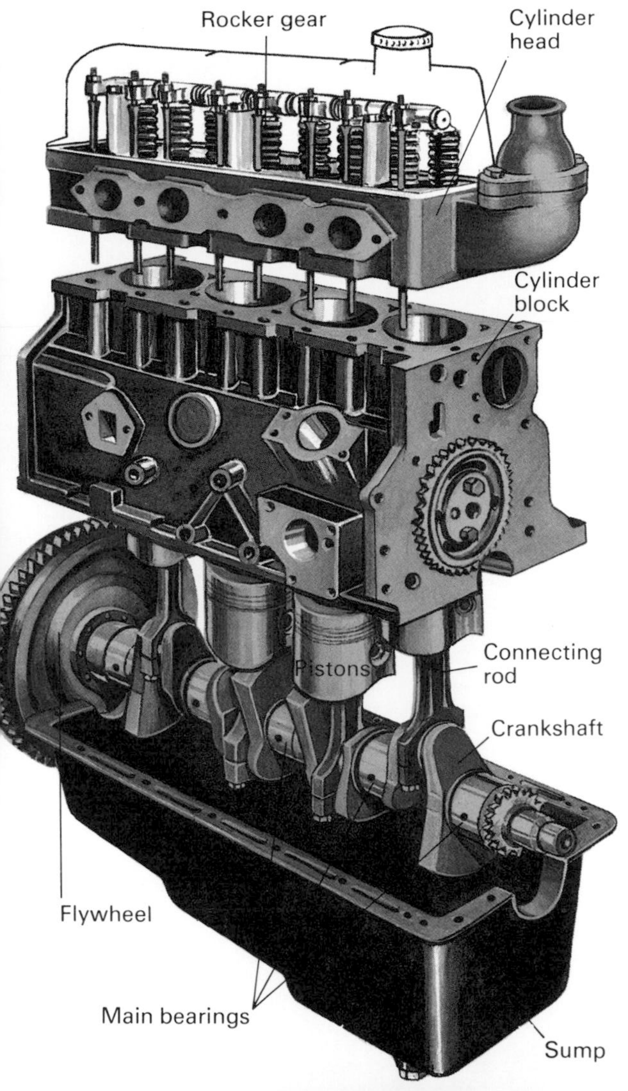

The ignition system provides the spark to ignite the fuel in the cylinders. It takes the low voltage of the car battery (12 volts) and boosts it to pulses of 15,000 volts or more. The pulses go to a distributor, which distributes them to sparking plugs in each cylinder, where sparks are produced. Most ignition systems consist of a coil (a kind of TRANSFORMER) and a contact-breaker. The contact-breaker constantly interrupts the battery current going to the coil, and this generates the pulses of high-voltage electricity in the coil. Electronic ignition systems, using transistors instead of contact-breakers, are now coming into more widespread use.

A cooling system is needed to rid the engine of the heat produced when the fuel burns, at temperatures of over 700°C. Most cars are cooled by water, which circulates through passages in the engine block. The hot water coming from the engine is itself cooled by cool air flowing through a radiator. Some cars and practically all motor-cycles, however, are air-cooled. Cool air flows past metal fins around the cylinders and carries away the heat.

The lubrication system also helps to carry away some heat. But its main job is to lubricate the parts that move within the engine with oil. The system pumps oil from a reservoir in the sump to rotating parts, such as the main bearings on the crankshaft. Without lubrication, such parts would soon heat up because of friction and seize, or lock solid.

The exhaust system removes the burnt gases from the cylinders and discharges them at the rear of the car. It also slows down the gases, which leave the engine at supersonic speeds. It does so by means of a muffler, or silencer.

Motor Cars 1

Below: A cutaway view of the Ford Sierra, a sleek five-seater family saloon with fine aerodynamic lines. It has independent suspension at the rear as well as at the front. This model is the 2·3 litre Ghia, which has a V6 engine developing 114 horsepower. Its top speed is 190 km per hour.

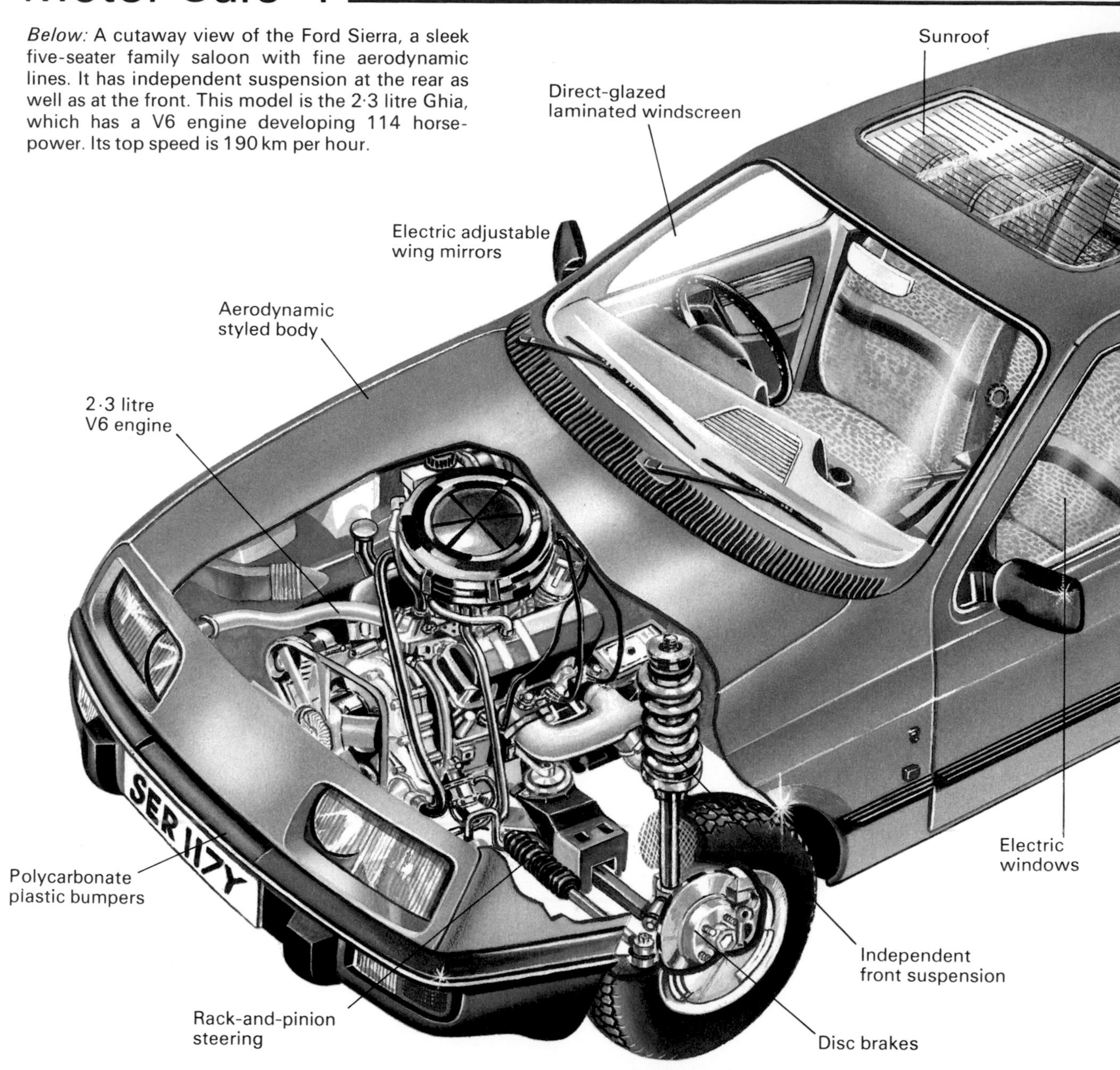

A century ago there were very few power-driven vehicles on the roads, apart from a few steam carriages. These were noisy, dirty and, because of their boiler, liable to explode! They were considered so dangerous that in Britain a man had to walk in front of them carrying a red flag.

In the late 1880s, however, two German engineers called Gottlieb Daimler and Karl Benz produced, independently, a new kind of vehicle. It was light and clean and powered by a petrol engine. It was the forerunner of the modern motor car, or automobile, which has since transformed society.

At first, skilled workers were required to build cars, and they remained quite expensive. Then in 1908 Henry Ford in the United States changed all that by setting up a production line to manufacture a single model – the famous Model T, or "Tin Lizzie". In 19 years of production an incredible 15 million Model Ts were made at a price the ordinary man in the street could afford. It revolutionized people's way of life, giving them the freedom to travel where and when they liked.

This was certainly true in the United States, which still has by far the largest number of cars – over 123 million of them. But the U.S.A. is no longer the world's biggest car manufacturer, having been recently overtaken by Japan. In 1982 Japan produced nearly 7 million cars, out of a total world production of about 50 million.

The modern car is a highly sophisticated machine, made up of as many as 14,000 different parts. Every year new models are introduced which offer some improvement on the previous ones. They vary from simple "people-carriers" like the Citroën 2CV; comfortable family saloons like the Toyota Carina; useful load-carriers like the Volvo 240 estate; lively sports coupés like the Chevrolet Camaro; luxury saloons like the Cadillac Eldorado; and matchless convertibles like the Rolls-Royce Corniche. There are now even "talking" cars, like the MG Maestro.

Left: To achieve efficient aerodynamic lines, car makers test their designs in wind tunnels. Here, a German-built Sierra is being evaluated in a wind tunnel at Cologne. Smoke is introduced into the air stream blown past the car so that the flow pattern can be easily seen and photographed.

Right: The sure-footed Audi Quattro, which has four-wheel drive. It also has a computer and voice synthesizer. The top model in the range has a top speed of 220 km per hour.

Motor Cars 2

The Body

The motor car of today bears little resemblance to the bone-shaking "horseless carriage" found on the roads at the beginning of this century. Most cars today are sleek-looking vehicles with a smooth and streamlined, or aerodynamic shape. With good body streamlining, less engine power is wasted in overcoming air resistance, or drag. Designers achieve good streamlining as a result of testing models of their cars in a wind-tunnel, just as aircraft designers do.

Most cars are made with an "all-in-one" body shell, consisting of a number of steel panels welded together. This type of construction lends itself to modern automated assembly methods (see page 65). It is also lighter than a separate chassis, as earlier cars had.

Manufacturers try to design cars to be as safe as possible, and carry out crash-tests to see if they can improve body design.

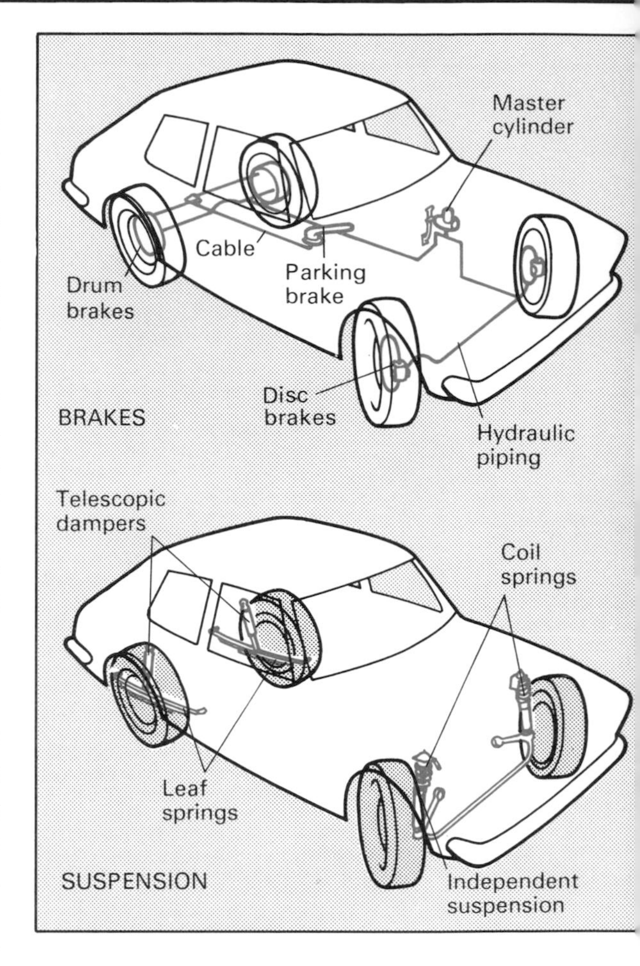

The Engine

Petrol engines still power most cars (see page 88). They are continually being improved to reduce the pollution they cause and to give them better fuel consumption, or more kilometres of motoring per litre of fuel. Driven carefully, small "hatchback" cars like the Talbot Samba can travel up to about 20 km on a litre of fuel.

To achieve better performance without unduly increasing fuel consumption, many cars are now being fitted with a turbocharger. This uses a fan to force more air into the engine. The fan is driven by a turbine in the car exhaust.

Many manufacturers are now offering their cars with diesel engines (see page 97) instead of petrol engines. Only in recent years have they designed diesels light enough for ordinary family saloons and with acceptable performance. Diesels are now available even for small hatchbacks like the Volkswagen Golf. The Golf Diesel is turbocharged and can travel at over 150 km per hour.

Computers are also finding their way into cars. They "supervise" electronic systems that control engine efficiency and work out such things as average speed and fuel consumption. Other electronic devices are also now being offered, including speech synthesizers, which scold the driver if he has left a door open or forgotten to put on his seat belt.

Looking to the future car manufacturers are experimenting with hydrogen-powered engines; hot-air engines like the STIRLING ENGINE; and novel steam engines. They are also looking into electric cars.

Below: Crash-testing cars at Renault's technical research centre at Lardy, near Paris. By such tests car designers investigate how well their vehicles would withstand collision on the roads. They use life-size dummies in the seats and record what happens to them during impact. Car designers try to ensure that the main passenger compartment remains rigid and undistorted during collisions.

The Drive

Until the 1960s most cars had rear-wheel drive. The engines were mounted in the front and drove the rear

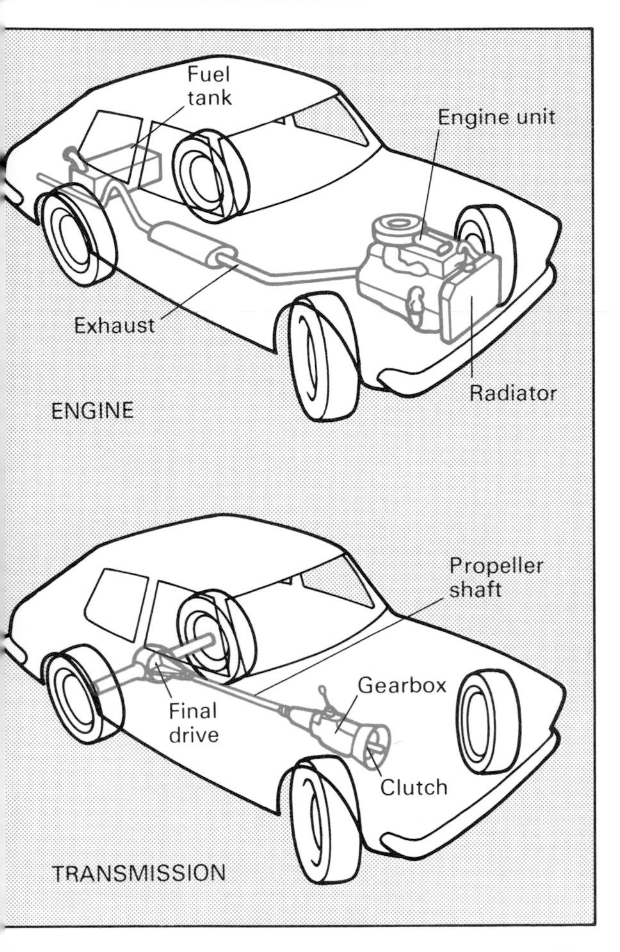

Above: Four of the main systems in a car. The engine produces power by burning petrol, stored in a tank at the rear. The transmission carries the power to the drive wheels, in this case at the rear. The brakes are applied by foot pedal and work hydraulically. Friction pads are forced against discs on the front wheels and drums on the rear wheels. In addition a mechanical handbrake works on the rear wheels. The suspension uses springs and dampers to cushion passengers from road shocks.

wheels via the transmission system. Today more and more cars, especially the smaller ones, have front-wheel drive. Their engines are mounted in the front and drive the front wheels.

The transmission system of a rear-wheel drive car consists of a clutch, gearbox, propeller shaft and final drive. In manual transmission, the clutch disconnects the engine from the rest of the system when the driver wants to change gear. The driver selects different gears in the gearbox to make the car travel at different speeds for the same engine speed. In an automatic transmission, the clutch and gearbox operate automatically, and gears change at certain engine speeds.

The propeller shaft carries motion from the gearbox to the final drive. The final drive changes the direction of motion in order to drive the rear wheels. It includes a set of gears ("differential") that allows the rear wheels to travel at different speeds around corners, since one wheel must travel further than the other.

A front-wheel drive car needs no propeller shaft, and incorporates engine, clutch, gearbox and final drive in a combined unit. This arrangement is very compact and gives more space inside the car. It also improves road-holding in slippery road conditions. Superior to front-wheel drive in this respect is four-wheel drive, in which all four wheels are driven. Four-wheel drive has long been used for cross-country vehicles, but is now becoming available on ordinary saloons. The four-wheel drive Audi Quattro was one of the first in this field.

The Suspension

Early cars were aptly named "bone-shakers" because they had hard cart springs and rigid axles front and rear. When the wheels hit a bump, the passengers felt the shock. Modern cars, however, have efficient suspension systems that cushion passengers from road shocks. In most cars the suspension consists of thick coil springs at the front and leaf springs at the rear. When the wheels hit a bump, the springs are compressed, which absorbs the shock. To prevent the body bouncing up and down too much, units called dampers, or shock absorbers are included. The suspension units always act independently at the front, and sometimes at the rear. Independent suspension greatly increases the smoothness of the ride.

A few cars have a totally different kind of suspension, which uses liquids and gases. It is called hydropneumatic suspension. It was pioneered as long ago as 1955 on Citroën's revolutionary DS19, and is still used on some Citroën models today. Some British Leyland models, including the Metro, have a hydropneumatic system called Hydrogas. In such a system springing is provided by the compression and expansion of a gas.

Speed Machines

Below: The Honda Sabre 750S superbike. It has a water-cooled V4 engine, six gears and shaft drive. With the engine running at over 9000 revolutions per minute, it has a top speed of over 200 km per hour.

Most car designers strive to achieve the greatest fuel economy from their engines, coupled with reasonable performance. An ordinary small family saloon like the Citroën GS, for example, has a 1·3 litre engine, with a power output of about 60 HORSEPOWER and a top speed of about 150 km per hour. It takes over 15 seconds to accelerate from 0–100 km per hour. It can travel about 15 km on a litre of fuel.

In contrast, some designers concentrate on performance rather than economy and come up with what can only be called "speed machines". The fastest include the Ferrari 400i, Llamborghini Countach and the Aston Martin V8. Their engines of about 5-litres deliver a power output of over 300 horsepower. The Llamborghini is the fastest, with a top speed of about 290 km per hour. They can accelerate from 0–100 km per hour in only about 6 seconds, and 0–150 km per hour in only about 12 seconds. But they can travel less than 4 km on a litre of fuel. They cost in the region of £35,000–£40,000.

Right: This view of an Alfa Romeo Formula 1 turbo shows the typical features of a Grand Prix racer – narrow cockpit, large rear "balloon" tyres, and aerofoils front and rear.

Bottom: Eddie Cheever's Renault RE30C, one of the turbocharged cars that have come to dominate Formula 1 Grand Prix racing in recent years.

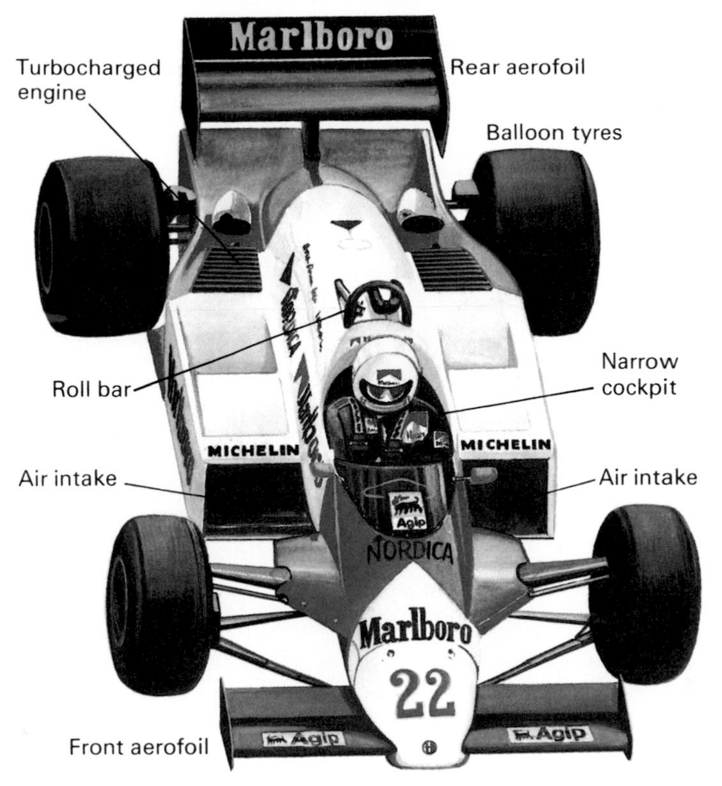

If you want that kind of performance for a fraction of the price, and reasonable fuel economy, you must go for two wheels, and get a superbike. One superbike is shown here – the Honda Sabre 750S. This can accelerate up to 150 km per hour from a standstill in just over 12 seconds. It is a machine fit for touring as well as racing. It has an engine of V design, which is water-cooled. And transmission is by shaft instead of the usual chain.

Some of the fastest cars of all are seen on the Grand Prix ("Big Price") motor-racing circuits. They are specially built for travelling at the highest speeds along the "straights" and bends of the twisting circuits. Along these straights the latest Formula 1 cars approach 340 km an hour.

Formula 1 cars may have ordinary engines of 3 litres, or turbocharged engines of $1\frac{1}{2}$ litres. The turbo-engined cars have a power output of more than 600 horsepower and began dominating Formula 1 racing in the early 1980s. Other noticeable features of Grand Prix racers are the aerofoils at the front and rear. These act in the reverse way to aeroplane wings. Air flowing over the foils pushes them *down*, and this helps keep the car on the track.

The ultimate in speed machines are those specials that are built to attack the world land speed record. Gary Gabelich achieved a speed of over 1000 km per hour in a three-wheeled rocket, *Blue Flame*, in 1970. And in the early 1980s this was being challenged by *Thrust 2*, a four-wheeler propelled by an aircraft jet engine. This was designed to travel at speeds up to the SOUND BARRIER – over 1200 km an hour.

Commercial Vehicles 1

The roads carry an extraordinary variety of motor vehicles – commercial vehicles as well as cars. The most common commercial vehicles are trucks, or lorries, and buses. But the term also applies to other vehicles that have more specialist use. They include fire-engines, mixer trucks, and crane-carriers, together with municipal service vehicles such as dust carts and gulley emptiers.

Some of the biggest vehicles are designed for "off-the-road" use. They include the dumper trucks used for shifting earth and rock during construction work. The gigantic Terex Titan dumper truck can carry a load of 310 tonnes and towers 17 metres high when tipping.

In most countries trucks carry a very large proportion of the goods, or freight, transported overland. They are much more flexible in operation than their main rivals, trains. Trains can move large loads efficiently over long distances, but they cannot collect and deliver goods door to door. Sometimes truck and train combine to offer a very efficient freight-carrying system. The freight is packed in containers or trailers, which are transferred between truck and train at railway termini.

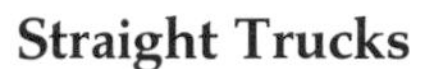

Straight Trucks

The smaller trucks are constructed with a single frame, or chassis unit. This is called a straight-truck design. Different bodies can be built on the chassis, for example, flat, dropside, tipper, box, tanker and coach. So a manufacturer can offer a variety of trucks for different uses at reasonable cost.

Most of these trucks have just two axles. The front axle, carrying a single wheel each side, is used for steering. The

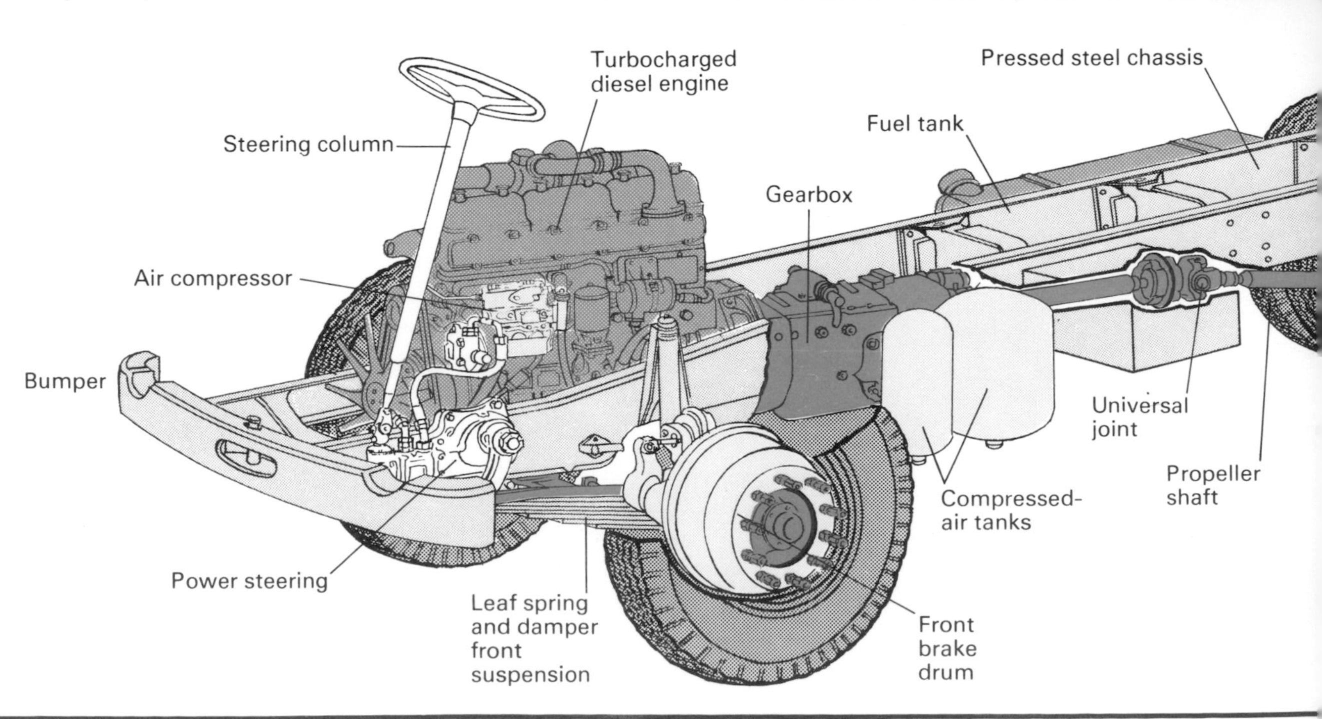

Left: Checking the stability of a double-decker bus on a tilt table. To allow for a generous safety margin in service, it has to withstand a tilt of nearly 30° without overbalancing.

Below: An off-the-road dumper truck carrying a full load. This rugged vehicle has four-wheel drive to give it extra traction in difficult ground conditions. It is also articulated for easier manoeuvring.

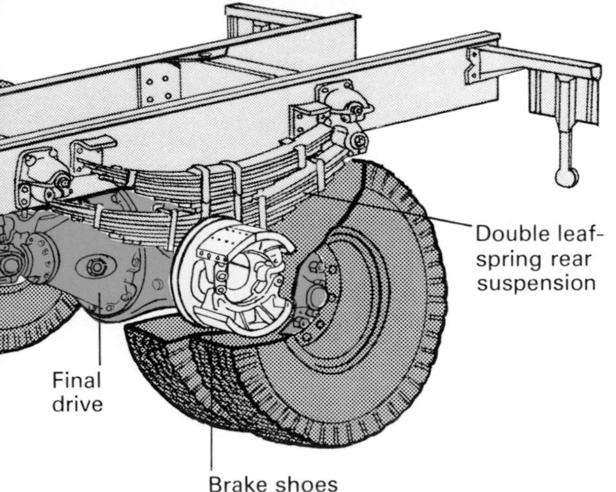

Above: The main features of a typical straight truck, on which various bodies can be built. It has a diesel engine and a multispeed gearbox, which drives the rear wheels. An engine-driven compressor supplies compressed air to power the brakes.

rear axle, carrying a single or a double wheel each side, provides the drive. Larger trucks may have three axles, with two drive axles at the rear; or four axles, with the two front axles for steering and two drive axles at the rear.

"Artics"

For hauling heavy loads, trucks are made articulated, with the vehicle effectively being split in two. The front part becomes a separate tractor with a short chassis, while the rear becomes a separate load-carrying semitrailer. Tractors may be four or six-wheelers with two, four or six-wheel drive.

At the rear of the tractor is mounted a coupling called a fifth wheel, or turntable. This connects with a coupling on the underside of the semitrailer. When coupled, tractor and semitrailer can swivel independently about the turntable, so that they can be easily manoeuvred. When uncoupled, the semitrailer is supported by legs at the front.

The articulated system is very flexible because a tractor can be used to haul a variety of semitrailers. It need not be idle while cargo is being loaded or unloaded from the semitrailers.

The Diesel Engine

A few trucks have petrol engines like cars, and some experimental ones have gas-turbine engines. But the engine used to power most trucks is the diesel, named (1892) after its inventor, the German engineer Rudolf Diesel. It is more rugged and more economical to run than a petrol engine.

The diesel engine is built in much the same way as a petrol engine (see page 88). It is a piston engine, which burns fuel in enclosed cylinders. The gases produced drive the pistons, which turn a crankshaft to provide rotary motion. The engine-operating cycle is repeated every four piston strokes, or movements. (Some diesel engines have two-stroke operating cycles. These are used particularly for powering ships.)

However, the diesel engine differs from the petrol engine in two important ways. One, it uses a different kind of fuel – light oil instead of petrol. Two, it burns this fuel in a different way, known as compression-ignition. This contrasts with spark-ignition in a petrol engine.

On the first stroke of the engine cycle, the piston moves down and draws air into the cylinder. On the second stroke the piston moves up and compresses the air into a tiny space. Because of this compression, the temperature of the air rises to over 500°C. Diesel oil is then injected into the hot air and immediately burns. The hot gases produced then force the piston down, producing power. When the piston moves back again, it pushes the burnt gases from the cylinder. The cycle then begins again.

Commercial Vehicles 2

Above: An articulated container truck. Containerization is now a favoured method of transporting goods by road, rail and sea. Inside standard-sized containers, goods can be easily transferred between truck, train and ship, simplifying handling and speeding delivery.

Below: "Doctor John", a powerful American tractor truck. It has the distinctive vertical exhaust pipes and silencers of commercial vehicles in the world's greatest trucking nation, where most long-distance freight is hauled by road.

Turbo Power

Diesel engines are made in a number of different designs and sizes. Six-cylinder engines with a power output of about 200 horsepower are widely used. In heavy trucks, engines with 8, 12 and 16 cylinders are used, which may have a power output of 500 horsepower or more.

In recent years power output has been increased by turbocharging. The turbo unit uses a compressor to force more air into the engine cylinders. It is driven by a turbine spun by the exhaust gases leaving the engine.

Transmission

To be able to haul a heavy load under different road conditions, a truck needs more gears than a car. Whereas a car generally has a gearbox with four gears, a truck may have one with thirty, though this is exceptional. The majority have five or six forward gears and one reverse in a main gearbox. In addition they may have a second, or "splitter" gearbox. This doubles the number of gears available. On many trucks the driver changes gear by hand, but others have automatic transmission.

Braking

A fully laden truck may weigh 50 tonnes or more. A great deal of power is required to move it – and to stop it. Ordinary hydraulic car brakes are of little use. So heavy

Above: This three-axle Super Metrobus of the Kowloon Motor Company runs in Hong Kong. It is 12 metres long and can carry up to 170 passengers – 110 seated and 60 standing. It has a 230 horsepower turbocharged diesel engine, automatic transmission and has air suspension as well as air brakes. Its body is made of aluminium.

trucks have compressed-air brakes. A compressor driven by the engine compresses air, which is stored in tanks. When the driver applies the brakes, compressed air flows to the brake cylinders and forces the brake linings against the wheel drums.

Articulated trucks have a "fail-safe" braking system, which applies brakes on the semitrailer should it become uncoupled. Most are also fitted with anti jack-knife equipment, which prevents articulated vehicles "jack-knifing". This is a dangerous condition in which the tractor swivels out of control around the articulated coupling.

Trucking in Comfort

The life of the long-distance truck driver is not as tough as it once was. The modern sleeper cab not only has bunks, but is often equipped as a small apartment. It has a cooker, refrigerator, hanging wardrobes and a wash basin. It is air-conditioned and insulated against noise and vibration, all of which helps keep the driver fresh and alert.

To improve the driver's working efficiency, the cab is equipped with a variety of instruments and devices that warn him, for example, if the brakes are wearing or if the load is shifting. The truck may even be equipped with closed-circuit television to look behind it, and with ultrasonic sensors to warn of nearby obstructions.

Below: By contrast the buses in Yosemite National Park in California are electrically powered so as not to pollute the natural environment. Visitors are encouraged to leave their cars and travel by bus, which is free.

Railways 1

Early last century there was only one way of travelling long distances – by stagecoach. That was tiring, uncomfortable and, because of highwaymen, dangerous. Then in 1830 came a great leap forward. George Stephenson in Britain built the first public, passenger railway to be operated entirely by steam locomotives – the Liverpool and Manchester Railway.

Its success led to a railway revolution that swept across the world. The railways provided essential transport for the Industrial Revolution that was taking place. They also helped the rapid settlement of new countries, particularly the United States and Canada. But above all they encouraged ordinary people, for the first time, to travel widely.

Earlier this century, however, the railways began to decline in importance in many countries because of competition from the car, lorry, bus and plane. But that decline is now ceasing. It is being realized that railways operated by modern locomotives can often provide a superior service. They can shift enormous loads and make very efficient use of fuel. They also cause less pollution and disturbance to the environment than their rivals. In many instances they provide a faster service between city centres than cars and planes.

Above: A "bullet train" on the Tokaido section of Japan's super-railway, the Shinkansen, with sacred Mount Fuji in the background. This section, from Tokyo to Osaka, has been operating since 1964 and handles over one million passengers a week.

The Locomotives

Until the 1950s practically all trains were hauled by steam locomotives. Then, gradually, railways in country after country began switching to diesel and electric locomotives. The reasons for the passing of steam locomotives are not hard to find. They were very inefficient. They could convert only about one-twentieth of the energy in their fuel (usually coal) into useful power and they created a great deal of smoke and soot.

Bottom left: A replica of George Stephenson's locomotive *Locomotion No 1*, built to celebrate the 150th anniversary of the opening in 1825 of the Stockton and Darlington Railway.

Only in a few countries in eastern Europe, South America, Africa and Asia are steam locomotives still in regular use. Elsewhere they can be seen only on a few lines run by steam preservation societies. But steam locomotives could one day make a comeback. Engineers in several countries are experimenting with new turbo-charged designs. They burn powdered coal efficiently, recycle their water, and cause little pollution.

Diesels and electric locomotives on the other hand, are highly efficient, quiet and clean. But to most people they lack the appeal of the old "steamers". Diesel locomotives use the same kind of engine as buses and trucks, which burn diesel oil (see page 97). But locomotives engines are very much more powerful. Some American diesels for hauling freight produce up to 6000 horsepower.

The power developed by the engine may be transmitted to the wheels in one of three ways. Most diesels have electric transmission, in which the engine is coupled to an electricity generator. This produces electricity, which is then led to electric motors on the driving axles. Diesel railcars and some of the smaller diesels used for shunting have a mechanical transmission, rather like that of a truck. A few other diesels have hydraulic transmission, in which power is transmitted by a kind of fluid coupling called a torque converter.

Faster than the diesels are the electric locomotives. A French electric TGV (Train à Grande Vitesse) currently holds the world rail speed record of 380 km an hour. Electric locomotives may be faster than diesels, but they are not so flexible. They can run only on track that has been electrified. And electrification is very expensive.

Below: One of France's TGVs, which operate the world's fastest service, between Paris and Lyons. Each trainset is made up of two power cars and eight trailer cars and can carry 386 seated passengers. It takes only about two hours to cover the 400-km journey.

The French TGVs and most other European electric trains pick up the current to power their motors from overhead power lines. They do so via a hinged contact arm, or pantograph attached to the roof of the locomotive. Most locomotives operate on very high voltage alternating current, usually 25,000 volts (100 times that of ordinary mains electricity). In a few countries, notably Britain, some electric lines are worked from a third rail. The locomotives pick up the current from a live rail that runs alongside the ordinary track.

Gas turbines are also now being used to power locomotives. The Union Pacific Railroad in the United States used massive gas-turbine locomotives for many years to haul freight trains through the Rockies. More compact units are now in use not only in the United States but also in Russia and in France, where they have achieved greatest success. The gas turbines work in much the same way as a jet engine (see page 108), but power is taken from the turbine rather than the jet exhaust. Usually the turbine drives a generator, and the electricity is fed to electric motors that drive the wheels.

Above: An electric locomotive of German Federal Railways (DB, Deutsche Bundesbahn) hauls a Trans-Europ Express (TEE) train through the spectacular scenery of the Rhine Valley.

Below: The severely streamlined nose of British Rail's Advanced Passenger Train (APT), which is designed to reach speeds up to 250 km per hour. Railway engineers achieve suitable shapes by testing models of their designs in a wind tunnel.

The Track

One reason the railways make efficient use of fuel is that there is low friction between the steel wheels of the train and the pair of steel rails that form the track they run on, called the permanent way. On modern high-speed lines, much of the track is now continuous-welded. It is made up of short lengths of rail welded together into sections several kilometres long. It makes for a much smoother and more silent ride. The parts are fixed to concrete sleepers embedded in very thick ballast, so as to withstand the pounding by high-speed trains.

Often for high-speed rail networks, the track is specially built, flat and straight, rather like motorways or expressways are in a road system. This has been done for the world's most successful high-speed network, the Shinkansen ("new trunk line") in Japan, on which the famous "bullet trains" run. On the Shinkansen average speeds of over 160 km per hour are common. On the TGV lines trains may run at speeds up to 260 km per hour.

In other countries, railway engineers have experimented with tilting train bodies so that high speeds can be maintained even on existing, curved track. Canada's turbo-powered LRC ("Light Rapid Comfortable") trains have a tilting body; so has Italian Railways' Pendolino. Experiments with tilting-body trains are also taking place in Britain (with the Advanced Passenger Train) and Spain.

Left: One of the latest "bullet trains", which travels on the Tohoku section of Japan's Shinkansen between Tokyo and Morioka to the north-east. The trains on the Tohoku Shinkansen have this attractive cream and green livery, contrasting with the blue and ivory of the Tokaido (Tokyo–Osaka) and Sanyo (Osaka–Hokata) Shinkansens. They also have more powerful motors to cope with the steeper gradients they encounter. The Tohoku Shinkansen opened in mid-1982, followed a few months later by the Joetsu Shinkansen to Niigata. By early 1983 the total Shinkansen network extended for more than 1800 km.

Below: Fiat's Pendolino, one of several tilting-body designs now being tested on the world's railways. The idea behind such designs is to achieve much higher speeds on existing curved track, without throwing the passengers off balance.

Signalling and Control

Modern methods of signalling and train control have also improved the efficiency of the railways. Easier-to-see coloured light signals have replaced the old semaphore-arm type. And these signals are no longer set by men in signal boxes every few kilometres along the track. They are operated from a traffic control centre, which may cover hundreds of kilometres of track.

At these centres the rail network covered is displayed in miniature on a panel. Operators can set any signals and track-changeover points within the network just by flicking a switch. Lights on the panel indicate the position of trains on the tracks. This is done by means of track circuits – electrical circuits which are triggered off when a train goes over them.

The high-speed TGV lines have no signals along the track. All traffic information is relayed to the train through track circuits and displayed in the driver's cab.

Mass Transit

It is not only inter-city services that are being improved. Rail traffic within cities is also developing fast. Travel by road in the cities is becoming slower and slower because of traffic congestion. So more and more underground railways, or subways, are being built.

One of the most recent is Hong Kong's mass-transit railway (MTR), which opened in 1980. "Mass-transit" is a good description because the railway is designed to carry two million passengers a day. This is about the same as the number carried on the London Underground, which is a very much larger system and was the first underground railway to open, in 1863. It is still the world's largest, with 420 km of track.

In most city mass-transit railway networks the routes include not only underground track, but also surface track and elevated track above road level. San Francisco's Bay Area Rapid Transit (BART) scheme is like this, with 30 km of tunnels, 40 km on the surface and 50 km above ground. Modern MTRs operate largely under automatic control.

Monorails

A totally different kind of elevated railway can be found in some cities. The passenger cars do not run on the usual two rails, but on a single rail, or monorail. The first successful monorail system, the Wuppertal Schwebebahn in Germany, has been running since 1901. Its cars are suspended from an arm attached to a kind of trolley that runs along an overhead rail. A suspension monorail can also be seen at Tokyo Zoo.

Tokyo also has another type of monorail, of straddle design, connecting with the airport. The passenger cars straddle a concrete beam that forms the track. It runs on

Above: In 1980 travel within the densely populated city of Hong Kong was revolutionized by the opening of the Mass-Transit Railway (MTR). Its 15 stations along 16 km of track serve nearly 5 million people.

Below: The straddle-type monorail that carries visitors around Disneyland near Los Angeles.

wheels on top of the beam and is driven by wheels that run along the sides of the beam. Straddle monorails take holiday-makers around Disneyland pleasure parks in the United States.

Flying Trains

Monorails also feature in designs for the trains of tomorrow. Railway engineers reckon that these trains will run at speeds of up to 800 km an hour! But they will not have wheels, because wheels cannot grip the track at this speed.

Instead, tomorrow's trains will float above the track. They may be air-cushion vehicles, like a hovercraft (see page 121), and glide on a cushion of air. The French Aérotrain has shown this to be a possibility. Or they may be suspended above the track by means of magnetism. This form of lift is called maglev, short for "magnetic levitation". Many people are already experimenting with maglev trains, and it is probable that the first one will run commercially on the Shinkansen.

Maglev depends on the simple principle that magnets can repel each other. The maglev train carries a very powerful magnet, made from SUPERCONDUCTORS. When it travels fast over a metal track, it makes the track magnetic too. The magnetic track repels the train's magnet and lifts up the train.

Both the hover and maglev trains would be propelled by a different kind of electric motor from the ones that propel today's electric locomotives. It is called a linear induction motor. An ordinary electric motor spins things round when electricity is passed through it. By contrast the linear induction motor moves things in a straight line.

Bottom right: The French Aérotrain, an experimental hovertrain that glides along a monorail track on a cushion of air. This jet-powered vehicle has reached a speed of over 400 km per hour. Other models have been electrically powered, by linear induction motor.

LONDON

Chapter 6

Transport by Sea and Air

Ships are the oldest form of man-made transport and, as freight carriers, are as important today as ever. Present-day passengers sail in ships mainly for pleasure rather than business because ships are so very slow compared with other means of transport. Faster sea-going craft, however, are now appearing on the scene for ferry service. They are the surface-skimming hydrofoils and hovercraft.

Most passengers travelling overseas prefer the superior speed of the aeroplane. The first flights were made by the Wright brothers on 17 December 1903. Some 80 years later streamlined planes like Concorde can whisk passengers across the Atlantic in less than four hours at the speed of a rifle bullet.

The recently built dry dock complex at Dubai, which has greatly expanded the facilities available for shipping. One of the Persian Gulf states, Dubai is a major centre for international trade in the Arab world.

Aircraft 1

Above: With its droop nose lowered, Concorde comes in to land. The only successful supersonic airliner, Concorde flies faster than a rifle bullet at a speed of about 2170 km an hour (twice the speed of sound, or Mach 2).

It is strange to think of something as big as an airliner travelling faster than a rifle bullet. But that is what Concorde does when it crosses the Atlantic Ocean in just three hours. Concorde is a supersonic airliner – one that can travel faster than the speed of sound (MACH 1).

Sharing the Atlantic route with Concorde are hundreds of other airliners, such as Boeing 747s and Douglas DC-10s. These are much slower, subsonic craft that travel at less than the speed of sound. However, they have the advantage that they can carry 400 or more passengers – three times the number Concorde can carry. The 747s are gigantic machines, nicknamed "jumbo-jets", with a take-off weight of over 300 tonnes, a wing-span of 60 metres and an overall length of 70 metres.

All Shapes and Sizes

The aircraft flying today come, literally, in all shapes and sizes. At the other end of the scale from the big airliners, there are the small private aircraft like the Piper Tomahawk and Beech Bonanza. There are even manpowered aircraft like the Gossamer Albatross, which crossed the English Channel in 1979. On the military side there are fighters even faster than Concorde, such as the Russian Mig-25, or "Foxbat", which exceeds Mach 3. Designs have been prepared for planes that travel at hypersonic speed – more than Mach 5 or five times the speed of sound.

In the immediate future, however, aircraft designers are concentrating mainly on designs that are more efficient and that in particular use less fuel. Boeing's new designs include the 767, which went into service in 1982. It is much smaller than the 747, carrying a maximum of 210 passengers, but it can carry them much more economically.

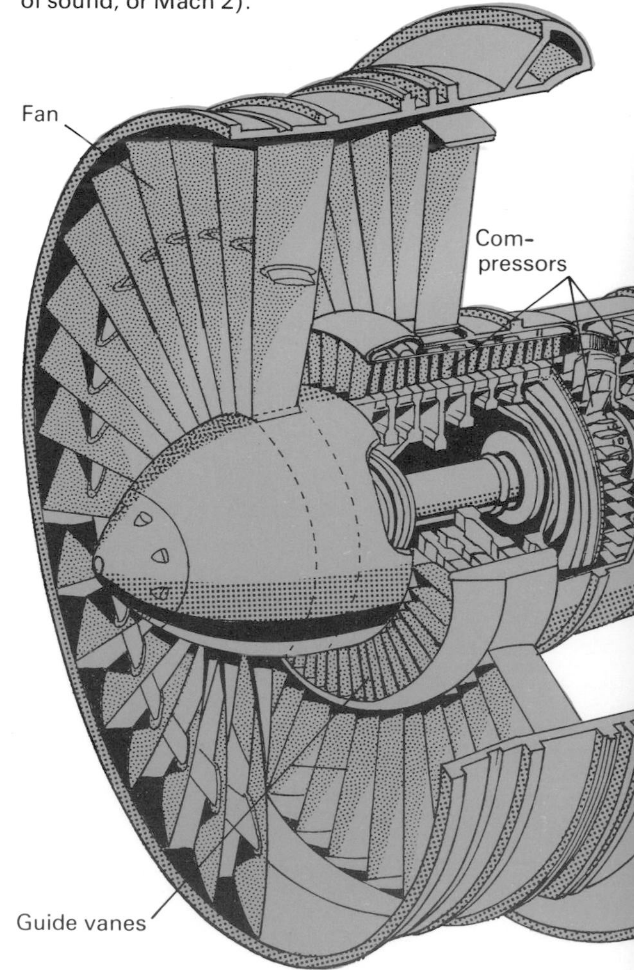

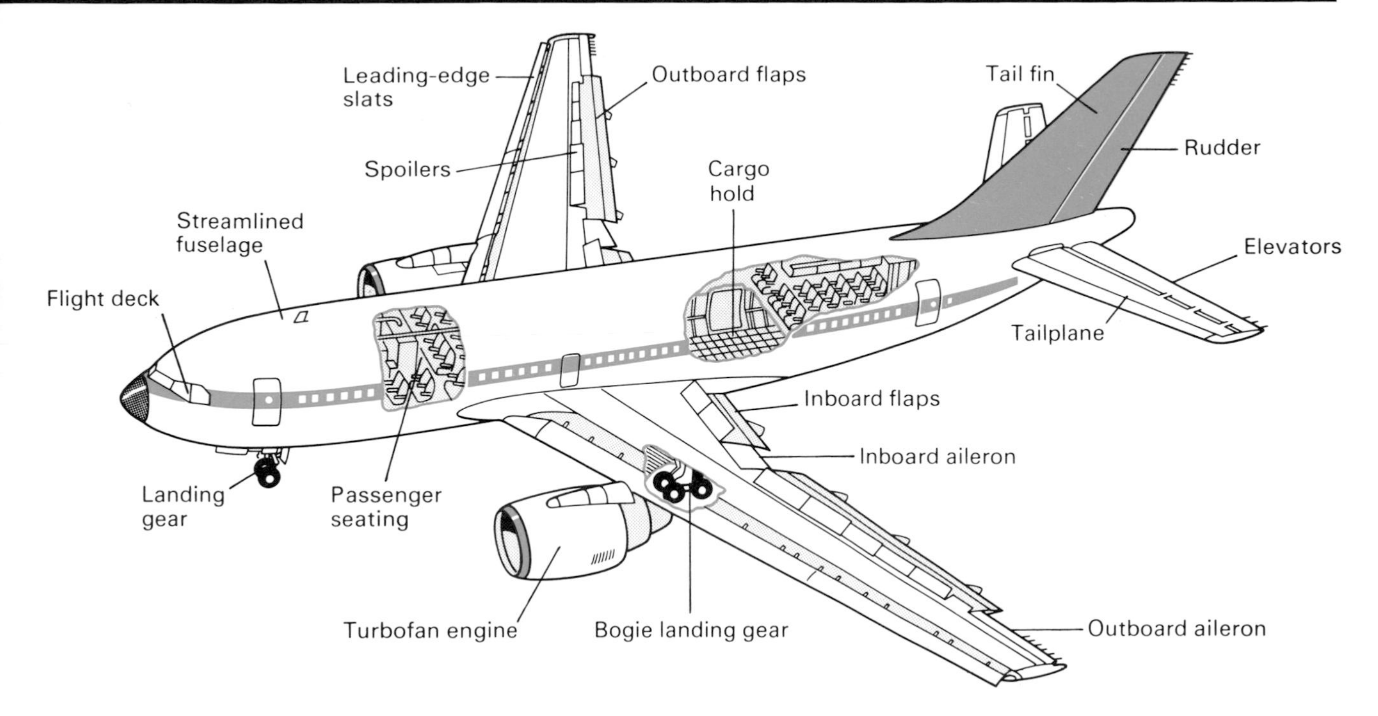

Above: The main features of a modern jet airliner, the European A300B Airbus. It has two powerful turbofan engines, which each develop a thrust of over 20,000 kilograms. It has relatively slim, "supercritical" wings, across which numerous slats, spoilers and flaps regulate the air flow.

Turbines
Tail cone
Combustion chamber
Concentric turbine/compressor shafts

Above: A modern turbofan engine, whose intake fan is typically more than 2 metres across. It has three turbine and three compressor stages.

It now competes with other fuel-efficient aircraft like the slightly larger European A310 Airbus.

Jet Power

Until the 1950s most planes were powered by piston engines, which worked in much the same way as car engines and burned petrol as fuel. The engines turned propellers, which developed the thrust to propel the planes through the air. Today, however, most planes have jet engines, or more accurately GAS-TURBINE engines.

Most fighter planes have turbojet engines. In the turbojet, fuel (kerosene) is burned in compressed air to produce hot gases, which escape from the rear in a high-speed jet. As the jet shoots backwards, the engine (and plane) is thrust forwards by REACTION. Most airliners, however, have turbofan engines, which are more economical. The turbofan has a big fan in front which forces air not only into the engine but around it as well. The "by-pass" air mixes with the hot gases coming from the engines to produce a more efficient propulsive jet.

Slower transport planes have turboprop, or propjet engines. In the turboprop, the jet exhaust spins a turbine which drives a propeller. Most thrust comes from the propeller, only a little from the jet exhaust. Turboprops make efficient use of fuel, but they cannot propel a plane at speeds above 650 km per hour because then the propeller tips start to travel at supersonic speed and set up shock waves. However, NASA have developed a new type of propeller called a propfan which can spin at very high speeds without causing shock. Propfan engines would be very economical to run and could be powering airliners and cargo planes by the 1990s.

Aircraft 2

The Wings

It is easy to understand how a balloon, which is lighter than air, rises from the ground. But how does a 300-tonne plane overcome gravity? The answer lies in the wings, and particularly in the shape of the wings. They have what is called an AEROFOIL cross-section, rounded at the top and flat underneath. When air flows past a wing of this shape, the wing tends to rise upwards. So, to take off from the ground, a plane has to travel faster and faster until the upward "lift" on its wings is greater than its weight. Then it rises into the air.

In low-speed planes the wings project more or less at right-angles from the fuselage. Most jet planes, however, have their wings "swept back", rather like an arrowhead. Such a design reduces the air resistance, or drag, on the wings as speed increases. To travel at supersonic speeds, the wings need to be sharply swept back. In some designs, including Concorde and the Saab Draken fighter, they merge into a triangular shape called a delta wing.

Some fighter planes, including the Panavia Tornado and the F-111, have "swing-wings", correctly known as variable-geometry wings. For low speeds, as at take-off and landing, the wings extend straight and give maximum lift. For high speeds the wings swing into a sharp swept-back position. For the future many other strange wing designs have been proposed, such as the forward-swept wing for fighters to make them faster, lighter and more manoeuvrable. Advanced civilian designs include the delta-wing Spanloader and the Ring-Wing (see page 112).

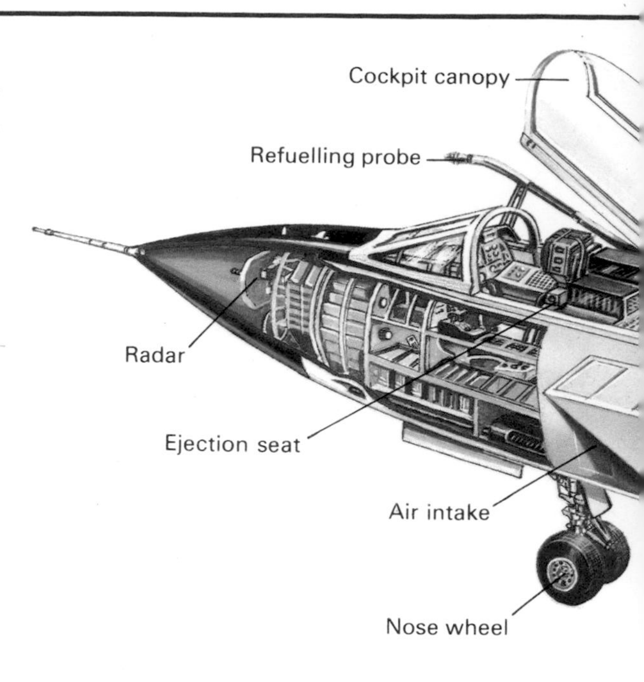

Above: The Panavia Tornado, a multirole combat aircraft now in service with European airforces. It is a co-operative venture between British, German and Swedish aerospace companies, and is a swing-wing design. At take-off and landing and for low-speed flight, the wings are outstretched in their forward position. This gives them maximum lift. For high-speed flight the wings swing back to form a near continuous surface with the tailplane.

Left: A scale model of Concorde being tested in a high-speed wind tunnel. Models are usually suspended upside-down to simplify calculations of the forces acting on them.

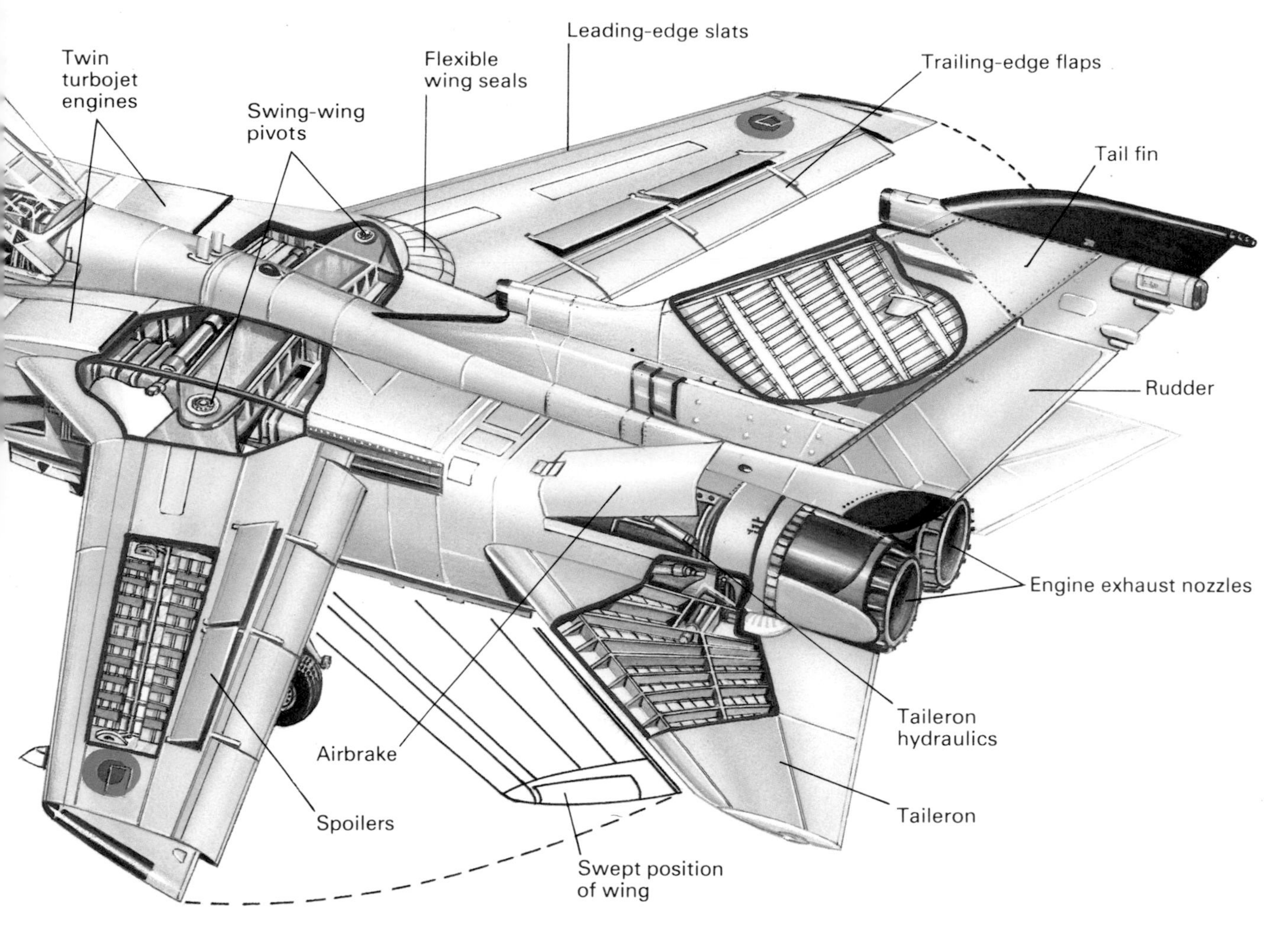

Right: Concorde's flight deck is crammed with instruments that monitor the operation of the plane's engines and flight systems.

Below: The A310 Airbus is one of the new breed of lighter weight, fuel-efficient airliners.

Aircraft 3

Design

All planes have the same basic parts. They have a main body, or fuselage, to carry passengers and cargo. They have wings, which support them in the air, and engines to propel them through the air. They also have a tail, consisting of a vertical tail fin and a horizontal tailplane. The main function of the tail assembly is to keep the plane steady, just like the flight feathers keep a dart steady. Control over the plane's movements in the air is exercised by moving hinged control surfaces at the rear of the wings (ailerons) and tail (elevators and rudder).

The way these basic parts are put together varies from plane to plane, depending on what it is to be used for, its operating range, cruising speed, and so on. As mentioned earlier there are several shapes of wing and types of engine for aircraft designers to choose from. They can place the engines in various positions, under the wings, in the tail, alongside the tail, or in any combination of these. To help them select a suitable design, extensive testing is carried out in a WIND TUNNEL. This is called aerodynamic testing.

First they build an exact scale model of their design and then suspend it in the wind tunnel. They note how the model behaves when air is blown past it. From this they can estimate how a full-size plane will behave when it is flying through the air. The aircraft designer may also carry out tests in a water tunnel. This is called hydrodynamic testing.

Construction

Wind- and water-tunnel testing help the designer to produce a shape for his plane that has the least air resistance, or drag. In construction, pains must also be taken to ensure that the outer surface of the plane is as smooth as possible, because a rough surface also causes drag.

Below: Some of the advanced designs for tomorrow's aircraft. The Spanloader and the Ring-wing are two designs put forward by the Lockheed Corporation. They could be flying before the end of the century. Forward-swept wing (FSW) prototype fighters are being built. And a NASA scissor-wing craft has already flown.

The Ring-Wing: A striking design in which the wings curve upwards and meet to form a ring shape. For the same lift, a ring-wing is lighter and causes less drag than an ordinary straight wing.

The Spanloader: A delta-winged transport plane which houses its cargo entirely within the wing. It should be able to carry a payload of 300 tonnes over 5000 km.

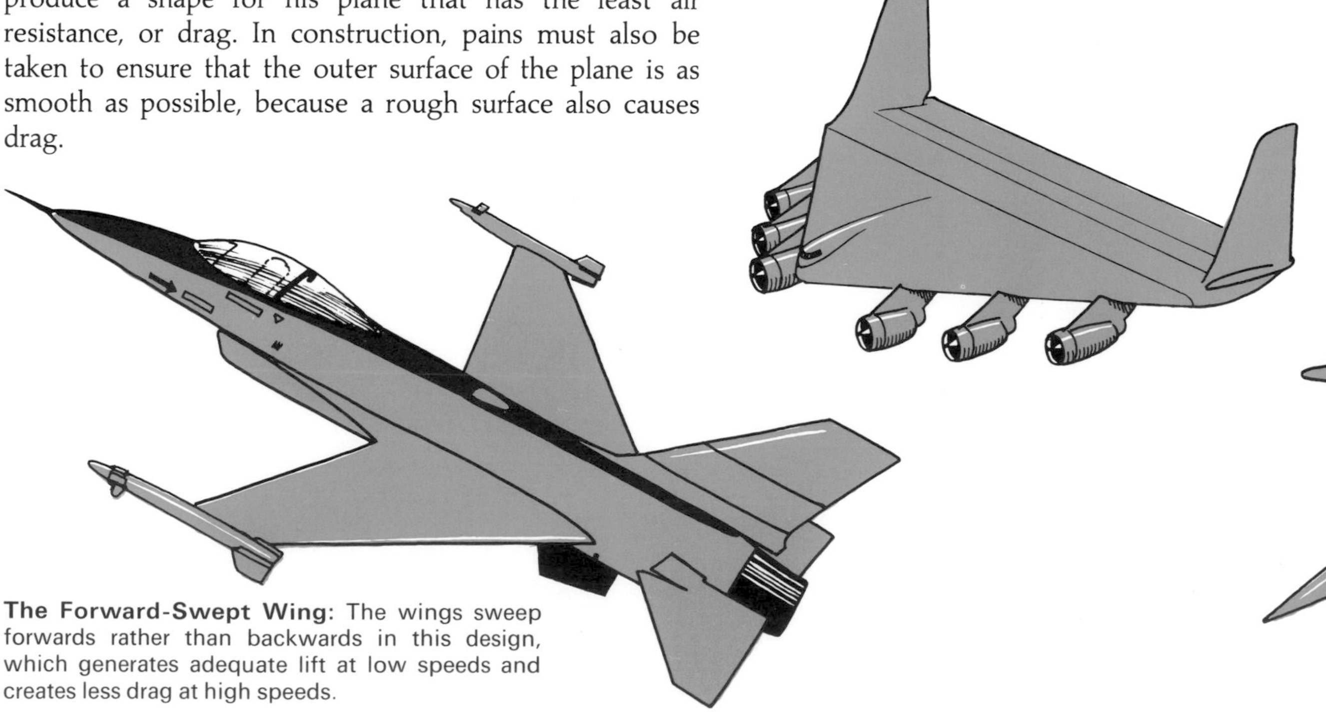

The Forward-Swept Wing: The wings sweep forwards rather than backwards in this design, which generates adequate lift at low speeds and creates less drag at high speeds.

Above: The production line for the European Airbus at Airbus Industrie's factory near Toulouse, France. Airbus production is an international enterprise, between the companies Aérospatiale (France), Deutsche Airbus (Germany), Fokker-VFW (the Netherlands), CASA (Spain) and British Aerospace. Each is responsible for the design and construction of different parts of the planes. British Aerospace, for example, builds the wings.

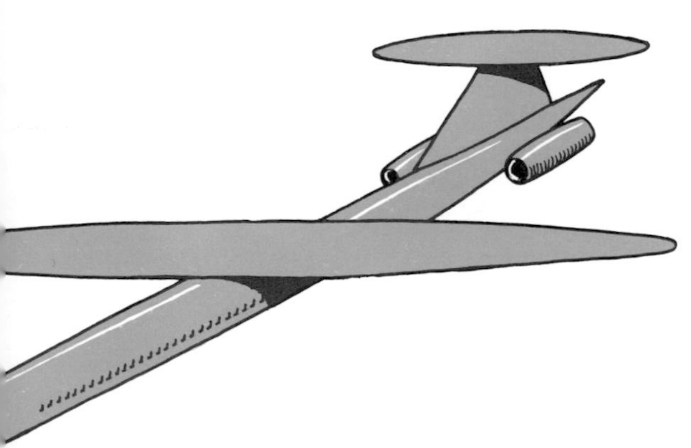

The Scissor-Wing: A variable-geometry plane with a difference. The wing extends at right-angles to the fuselage at take-off. Then it swivels on a central pivot, rather like a pair of scissors, so that half the wing is swept forwards, half is swept back.

The structure of the plane, or airframe, must at the same time be as light as possible, yet strong enough to withstand the forces that occur in flight. Just imagine what these forces are in a jumbo-jet: for example, the weight in the fuselage of 400 people and the thrust on the wings of four powerful engines.

Fortunately, there are modern alloys of aluminium, such as duralumin, which do combine lightness with strength. The main "skin" of a plane generally consists of sheets of aluminium alloy supported by a framework of spars, ribs and stiffeners. When extra strength is required, the skin is bonded to an aluminium honeycomb or sandwich; or the skin and supporting framework may be machined out of solid metal, which gives exceptional strength. In areas subjected to high temperatures, such as around the jet engines, stainless steel or titanium are used.

The way ahead for construction, of at least some designs, may well be in another direction, however – in the field of synthetic resins and plastics. Epoxy-resin adhesives are already widely used in aircraft construction to bond metal surfaces together. This is stronger and lighter than riveting, for example. But one plane has been built almost entirely from plastic materials, or composites. It is the Lear Fan, a small business plane with a single propeller at the rear, which first flew in 1981. Its entire body is made from epoxy resin, reinforced with CARBON FIBRE. It is much lighter, yet as strong, as similar size planes, and uses much less fuel.

VTOL

One drawback about most aeroplanes is that they require a long runway – up to 3 km for a heavy plane – for take-off and landing. Some aircraft, however, are designed to take-off and land straight up and down. They are called VTOL (vertical take-off and landing) craft. A successful VTOL fixed-wing aircraft is the British Harrier fighter. It is now in service in various countries, either in its ordinary air-force version, or as the modified Sea Harrier for operating from aircraft carriers.

The Harrier achieves vertical operation by what is termed vectored thrust. This means that it can change the direction in which the exhaust jet from its engine points. This is done by means of swivelling nozzles. For take-off the nozzles are angled so as to deflect the jet exhaust downwards, and the plane rises vertically. In the air, the nozzles are gradually swivelled to deflect the exhaust backwards, and the plane flies normally. The reverse happens on landing.

To keep it steady during the change from vertical to level flight, the Harrier is fitted with a reaction control system, rather like that on a spacecraft. This "fires" jets of high-pressure air from nozzles in the nose and tail and on the wing-tips to steady the craft.

Helicopters

The most successful VTOL craft, however, are helicopters, which are the most versatile of all flying machines. They can take-off and land vertically, hover, and fly forwards, sideways and backwards with equal ease. They can operate in any terrain, from mountaintop to sand dune, provided there is a small clear space.

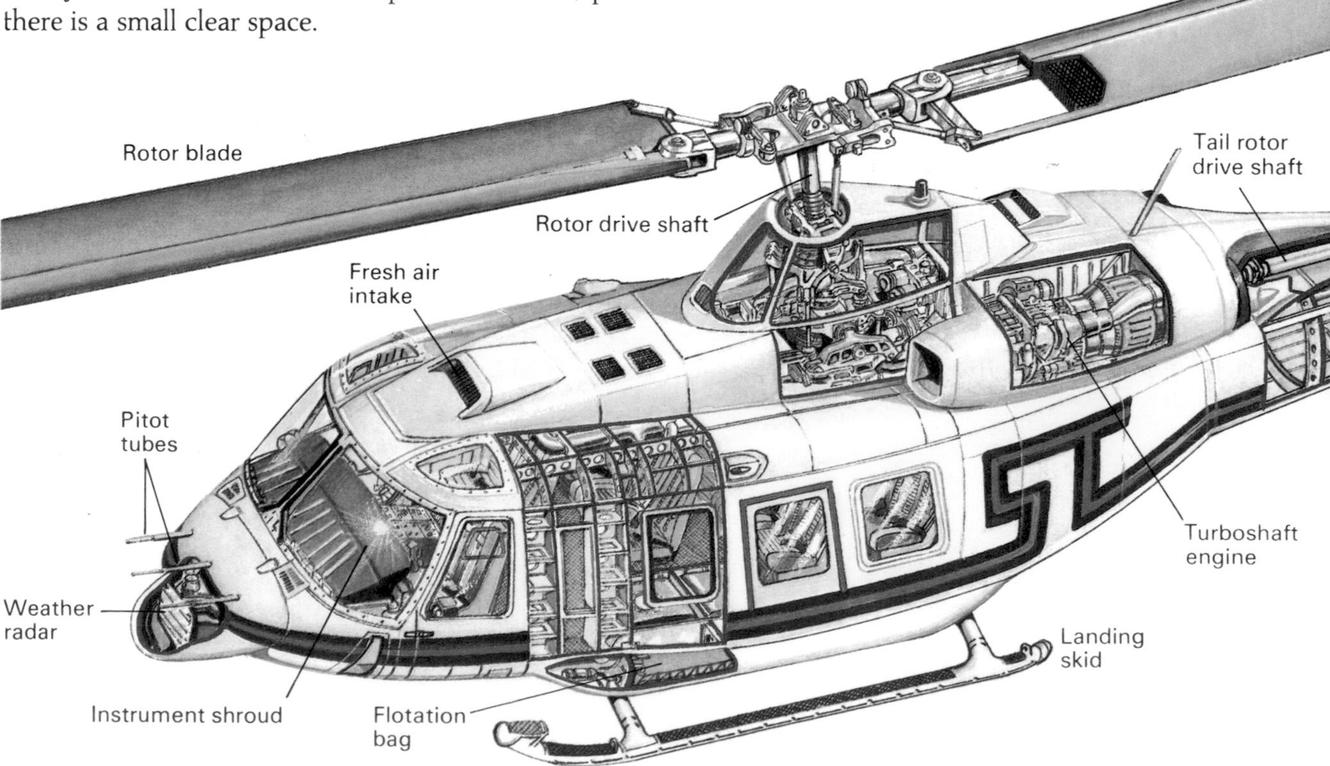

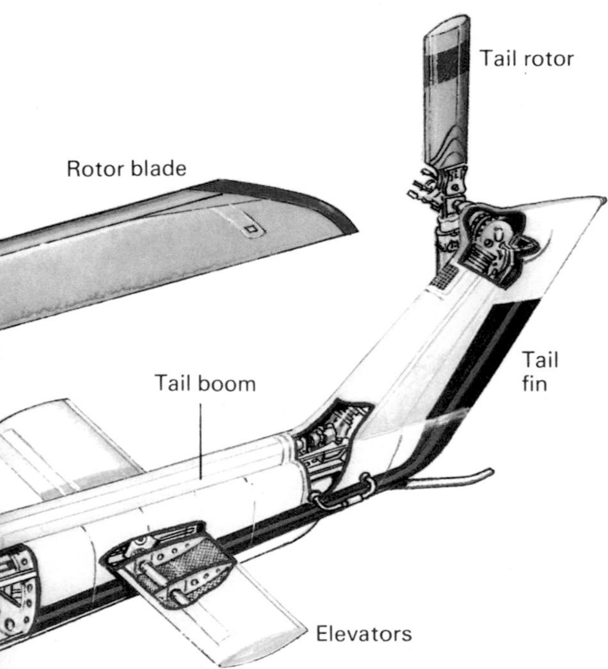

Above: The biggest commercial helicopter produced by Bell Aerospace, the 214ST, which went into service early in 1982. It seats 20 people, including the pilot. Its two gas-turbine engines give it a cruising speed of about 240 km per hour. Its twin-bladed rotors measure nearly 16 metres long.

Top: A Boeing Chinook helicopter used for transporting personnel to the North Sea oil rigs. It can seat more than 30 people.

Top left: The Goodyear airship *Europa*. Like all modern airships, it is a helium-filled blimp.

Centre left: The incomparable Harrier VTOL fighter, which achieves vertical take-off by swivelling ducts in the exhaust stream from its engine.

The military in particular have found abundant uses for their talents, as transports for carrying assault troops and equipment, and as gunships. Their ability to hover also makes them ideal for air-sea rescue operations. Police patrols use them; construction companies use them; businessmen use them; tourists use them; in Australia even farmers use them, for rounding up sheep and cattle.

The most important part of a helicopter is the rotor on top of the body. It consists of two or more narrow blades which have the AEROFOIL shape of a plane's wing. When the blades travel through the air, they provide lift in much the same way as an ordinary wing. By suitably angling the blades as they rotate, they can be made to propel the helicopter as well as support it in the air. The rotor is driven by shafts from a petrol engine or a gas-turbine engine.

When the blades turn, they tend to make the body of the helicopter turn too, but in the opposite direction. To prevent this happening, a helicopter has a small propeller mounted on the tail, facing sideways. It produces a sideways thrust to balance that of the body. Some of the bigger helicopters, such as the Boeing Vertol and Chinook, have two rotors – one at the front, one at the rear. They rotate in opposite directions, which means that they have no net effect on the body.

There are a few drawbacks to the helicopter. It is very noisy, because of its swivelling blades. It is potentially unstable; if a rotor breaks, then it immediately goes out of control. It is slow compared with a plane, with a typical cruising speed of less than 300 km per hour. And it has limited lifting capacity.

Airships

There is another type of VTOL craft that has been with us for much longer – the airship. Airships pioneered long-distance aviation in the early decades of this century. They included the famous German Zeppelins, such as the *Hindenburg*, a colossal cigar-shaped craft 247 metres long. In 1937, a year after its completion, it burst into flames while landing in the United States. This and similar tragedies ended airship construction, until quite recently.

Today, there is increasing interest in airships now that conventional aircraft have become so expensive to build and to operate. It was perhaps a sign of the times in 1982 when the Greek company Interport Marine ordered three airships from the British company Airship Industries, who are in the forefront of modern airship design. The airships are for carrying tourists between the Greek islands. In all modern airships, the light gas helium provides the lift. Early airships were dangerous because they were filled with hydrogen, which is highly flammable and potentially explosive.

Ships 1

Men first took to the seas in ships over 5000 years ago, travelling to distant countries and trading and fighting with people in foreign lands. They have been travelling, trading and fighting in ships ever since. Passengers have long ceased to use the ship as a major form of transport, preferring the speed of air travel. But as far as overseas trade is concerned, the ship still reigns supreme. Ships involved in trade between countries are called merchant ships, as opposed to naval ships designed for fighting.

At any time of the day or night tens of thousands of merchant ships, in fair weather and foul, are sailing on the high seas. They vary widely in size and in the type of cargo they carry. Some are simple coasters, maybe only about 50 metres long, which carry general cargo on short journeys from port to port. Others are giant tanker vessels transporting crude oil from the Middle East oilfields.

Oil tankers are by far the largest vessels afloat. The *Seawise Giant* completed in 1981, is a staggering 459 metres long and 69 metres across. It has a cargo capacity of over 500,000 tonnes.

Above: Container ships provide a very efficient method of carrying cargo. Goods are carried in standard-sized containers, which can be quickly loaded and unloaded at the docks.

Below: An oil tanker of typical design, with its smokestack and superstructure well aft. Like many modern vessels, it uses satellite communications via the global Inmarsat network, established in 1982.

Shipbuilding

For centuries shipbuilding has been one of the world's greatest heavy industries, although it has declined in importance in recent years. Japan is now by far the biggest shipbuilding nation, building nearly as many ships as all the other nations put together. Japan also has the largest merchant shipping fleet, of over 9000 vessels.

Until the mid-1800s all ships were built of wood. Today practically every ship is built of steel plates welded together. In the shipyards individual plates are first welded into larger sections, and then these sections are welded together in position on the ship's hull. Recently, some ships have been built of other materials, however. Some naval mine-sweepers are constructed of glass-reinforced plastic (GRP). A few ships have even been built of thin reinforced concrete, or ferrocement.

Every type of ship has a different hull design, but they all have certain features in common. They have a double-bottom – this means that the bottom is made up of double steel plates with a space in between. This is a safety measure in case the bottom gets holed. Horizontal plates at various levels form the decks of the ship. These are divided up into compartments by vertical plates called bulkheads. This construction strengthens the hull and is another safety feature because individual compartments can be sealed off if they become flooded.

Cargo ships in general do not carry much super-structure – the part of the hull above main deck level. There is the navigation bridge, or wheelhouse, and crew accommodation. And usually there is the smokestack, which leads from the ship's boilers and engines.

Below: The Japanese oil tanker *Shin Aitoku Maru*, launched in 1980. It was the first modern commercial vessel to use sail assistance.

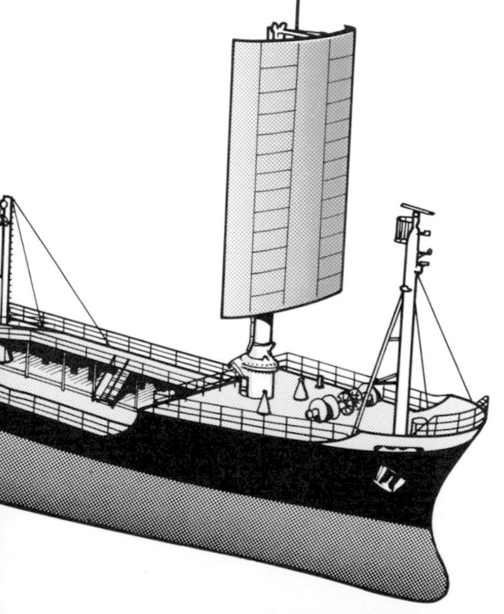

Ship Propulsion

Since the mid-1800s most large ships have been powered by steam engines and propelled by means of a propeller beneath the stern. The steam "engine" used today is not a piston engine as it once was, but a turbine similar to those used in power stations. Steam produced in a boiler is fed to the turbine and spins it round. The turbine is connected through a gearbox to a shaft that carries the ship's propeller, and spins this round too.

In most ships, the boiler is heated by burning fuel oil. In a few it is heated by a nuclear reactor (see page 18). The largest nuclear-powered vessels are the giant American aircraft-carriers *Nimitz* and *Dwight D. Eisenhower*. A few nuclear-powered merchant ships have been built, such as the German vessel *Otto Hahn* and the American vessel *Manhattan*, but they have not proved to be economical. But the largest group of nuclear-powered vessels are missile-carrying submarines. Russia and the United States have many nuclear submarines. The great advantage of a nuclear submarine from a military point of view is that it can remain submerged for months at a time.

The other main method of powering ships is by diesel engine. These are similar to the ones used to power trucks (page 97), but are very much bigger. Diesels now power most small ships, such as coasters and trawlers, and also many larger ones. Gas-turbine engines are fitted to some naval craft, notably the Royal Navy's aircraft carrier *Invincible*. But they are uneconomic for merchant ships.

For future propulsion some ship designers are looking to the past – to sails. The great attraction of sails is that they harness the power blowing in the wind, which is free. Already the Japanese have built a sail-assisted vessel, called the *Shin Aitoku Maru*. This uses two large sails controlled by computer to add to the engine power. The ship uses much less fuel than comparable ships of the same size. Other designers reckon that fully rigged "tall ships" similar to the tea clippers like *Cutty Sark* may soon return.

Bulk Carriers

Some of the biggest vessels afloat are bulk carriers, which usually have their superstructure well aft (at the rear) and a single deck with the cargo space beneath. The cargo may be coal, ore, cement, sugar or grain.

The oil tanker is a special type of bulk carrier, in which the oil is carried in individual tanks, to prevent the oil swilling back and forth too much, which would soon cause the ship to capsize. As mentioned earlier the biggest tankers are several hundred metres long. It is economical to build such long ships, because the longer the ship, the less power is required per metre to propel it. But long ships are very difficult to manoeuvre. Fully laden tankers can take up to a kilometre to stop.

Above: The *Norway*, formerly the *France*, is the longest (316 metres) liner ever built.

Opposite: An aerial view of the North Fleet Hope Container Terminal at Tilbury, near London.

Below: Bristling with antennae, the USS *Vandenberg* is a floating tracking station.

Roll on/Roll off

In cargo vessels of traditional design goods are loaded and unloaded vertically through hatches in the deck by means of cranes and derricks. This is a slow process. Many of the latest vessels are therefore designed for loading and unloading horizontally, as car ferries do. In this Ro-Ro (roll on, roll off) method, cargo can be handled swiftly and easily using vehicles such as fork-lift trucks.

Cargo handling has also been made easier by the introduction of the container system. Items of cargo are not handled individually, which is awkward and time consuming, but are packed in standard-sized containers (6 or 12 metres long and 2·4 metres square). Specialist handling equipment then carries out the loading and unloading.

A container ship has what is called a cellular construction. The holds are divided vertically into cells, in which the containers are stacked like bricks. Cellular container ships have several times the cargo-carrying capacity of ordinary cargo ships.

Float on/Float off

The idea of the container is taken a step further in the LASH (Lighter Aboard Ship) system. Here cargo is packed into standard-sized barges, which are towed out to the ship and then hauled on board by crane.

The system has been further developed into what is called flo-flo (float on, float off) operation. The ship is designed rather like a floating dock and by filling its ballast tanks sinks down into the water. After the barges have been floated in, the ship empties its ballast tanks and rises, lifting the barges with it. The Japanese again pioneered this system with the vessel *Mammoth Oak*.

RESOLUTION BAY
LONDON

Hydrofoils and Hovercraft

The great drawback about ships is that they are very slow compared with the other main forms of transport – the car, the plane and the train. Even fast ocean liners can travel at only about 35 knots (about 65 km an hour). Boats are slowed down by the tremendous resistance, or drag of the water on their hull. The inventor of the telephone, Alexander Graham Bell, was one of the first people to come up with a solution to the problem in 1918. He built a hydrofoil boat.

Hydrofoil boats are fitted with underwater wings, or hydrofoils. When they travel through the water, the wings provide lift in much the same way that a plane's wings provide lift in the air (see page 110). They lift the boat hull clear of the water, which means that it no longer experiences great drag. Its engines can therefore propel it at much higher speeds – to 80 km an hour or more.

Today large numbers of hydrofoil boats operate in some 50 countries around the world and carry tens of millions of passengers every year. Russia has by far the largest hydrofoil fleet. Many of the large hydrofoil craft outside Russia are built by the Swiss company Supramar. Their craft are fitted with V-shape, or surface-piercing hydrofoils. This design is more suited to inland waterways than to the open sea. For rough water submerged horizontal foils are better.

The most advanced hydrofoil craft, built by Boeing, has submerged foils and is fitted with an automatic control system to keep it stable in all sea conditions. It is called the Jetfoil, because it is propelled by water-jet engines. Jetfoils first entered service in 1975 between Hong Kong and Macau. Since then they have seen service in Hawaii and across the English Channel, notably between Dover and Ostend and Newhaven and Dieppe.

Above: The Boeing Jetfoil, "flying" on its submerged hydrofoils. It is driven by water-jet propulsion. Water is taken in through a duct in the rear foil assembly, and is pumped at high pressure out through nozzles at the base of the hull. The pumps are driven by gas-turbine engines. The craft has a fully automatic control system. It has gyroscopes and other sensors front and rear that instantly detect changes in the craft's attitude and direction and the wave height. This information is fed to a computer, which adjusts the position of the foils so as to keep the craft stable and on course.

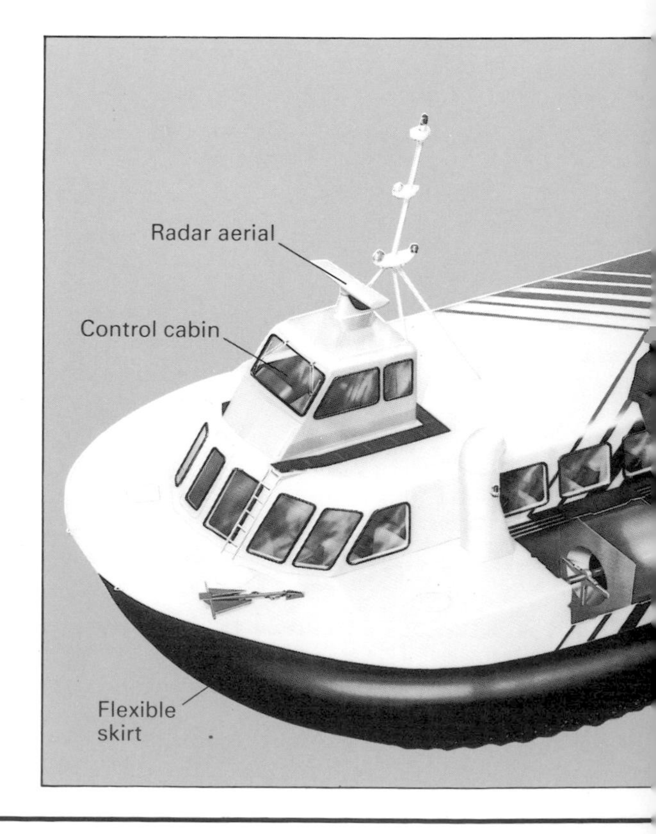

Above: A Russian hydrofoil craft of the Comet class. It has double V-shaped hydrofoils of the surface-piercing type.

Top right: Princess Anne, one of the successful SRN Super 4 hovercraft that provide a speedy car ferry service across the English Channel.

Below: The AP1-88 is one of a new breed of fuel-efficient hovercraft now entering passenger ferry service. It is powered by four diesel engines – two drive the shrouded propellers and two drive the lift fans.

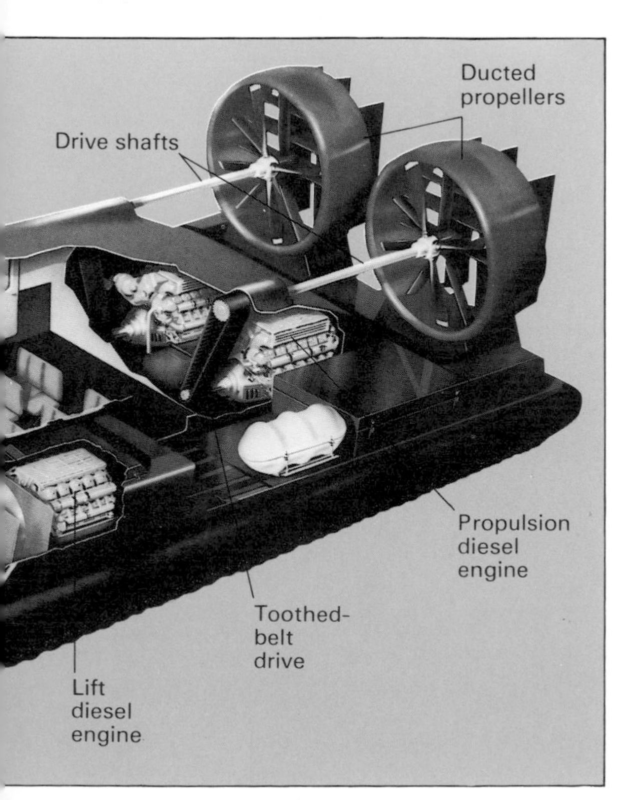

The Hovercraft

Even faster than the hydrofoil is the hovercraft. This is the British name for a machine called an air-cushion vehicle, which works on a principle discovered in 1955 by British engineer Christopher Cockerell. Many other applications of the air-cushion principle have been found, from "hover" lawn mowers to "hover" trains (see page 105).

The hovercraft looks nothing like an ordinary boat, but rather like a huge rubber dinghy. It does not need the typical hull of a boat, because it rides almost completely out of the water on a cushion of air. Freed from the drag of the water, it can thus move very fast.

Hovercraft have proved useful for the military as coastal patrol vessels and troop carriers, because they can travel with equal ease over ordinary roads, rough tracks and swamps. They have been very successful as car ferries.

Some of the largest hovercraft ferries operate across the English Channel, including the Sedam 500 and the SRN4. The SRN4 service has been particularly successful. The Super 4 version is nearly 57 metres long and can carry up to 420 passengers and 60 cars. It can travel at speeds up to 65 knots, or about 120 km per hour. The American navy has a large air-cushion craft called a surface-effect ship, which has travelled at over 90 knots, 170 km per hour.

In large hovercraft, like the SRN4, gas-turbine engines provide the power to lift and propel the craft. For lift, the engines drive powerful fans that force air beneath the craft to form a cushion. To prevent the air leaking away too quickly, the bottom of the craft has a flexible "skirt".

For propulsion, most hovercraft use propellers. They are usually backward-facing, or "pushing" propellers. They can be swivelled to help steer the craft, in conjunction with rudders on tail fins at the rear.

TVS

Chapter 7

Communi-cations

Ours is an age of "instant" communications. We can pick up the telephone and speak to someone in a distant city. We can switch on the television and watch events happening live practically anywhere in the world – and even on other worlds. We can send photographs by wire and business data by satellite. We can make home video films and listen to near-perfect recordings on digital discs.

Radio, television, teletext, video cameras and recorders, digital discs, facsimile machines, telex, teletypewriters, word processors, electronic mail – these are just some of the electronic miracles that are continuing the revolution in mass communications that started with the development of the mechanical printing press some 500 years ago.

Behind the scenes in a television studio during the transmission of the weather forecast. Note the use of two cameras – one filming the forecaster (right) and the other, the weather chart.

Printing 1

In about 1450 Johannes Gutenberg in Germany started a revolution in communications by developing the mechanical printing press. Over 500 years later the printed word in the form of books and newspapers is still the major method of storing and spreading knowledge. But this situation may not continue for long.

Sales of newspapers have already suffered heavily from competition from the television, which can report news virtually as it happens and, via satellites, from anywhere in the world. Other forms of data and information are no longer kept on paper on files or in books but are stored, much reduced, on MICROFILM, or electronically on tapes or discs in computer databanks.

Electronic gadgetry has, in fact, streamlined the whole printing operation, in the office as well as in the press room. We can now find electronic typewriters, typesetters, colour separators and even platemakers.

Above: Looking from the wet end towards the dry end of a Fourdrinier papermaking machine. At the wet end watery pulp mix is poured onto a broad wire-mesh belt, and the water gradually drains, or is sucked away. The damp paper "web" that forms is then dried and pressed by heated rollers.

The Electronic Typewriter

The electronic typewriter is rapidly superseding the electric typewriter in the office. The reasons are obvious. The most advanced electric typewriters, which have the type characters on a "golf ball", are very fast but also expensive. And they have some 2500 moving parts.

The electronic typewriter on the other hand has only about 100 moving parts. Instead of mechanical levers, swivel joints and springs, it has electronic sensors and is controlled by silicon chips (see page 144). The type characters are located on flexible "stalks" on a "daisy wheel". The daisy wheel can be changed in seconds to provide an alternative typeface. Furthermore, electronic typewriters are quiet in operation and, compared with the "golf-ball" type, very cheap.

Many electronic typewriters have a built-in memory

Right: Electronic typewriters use plastic daisy wheels like this to carry the characters of type. They are cheap, light and easy to change.

Left: Word processors in use. Between the two machines is a printer, which produces printed copies of the letters being typed. This particular type of word processor stores information on twin floppy disks in built-in units. Each disk can store over 300,000 characters of information, or about 50,000 words.

Right: An electronic typewriter is connected with a microcomputer to form a powerful word-processing system. The typewriter can also be used as an output printer for the microcomputer. But both microcomputer and electronic typewriter can still be used independently when required.

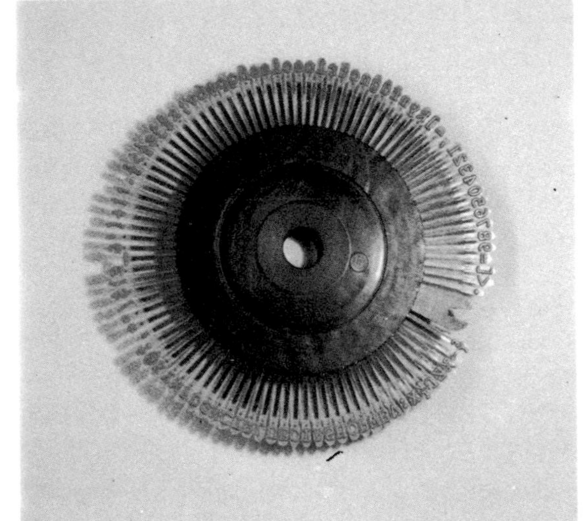

which can store simple information, including often used phrases such as "Dear Sir" and "Yours faithfully". They can also "remember" one or more of the previous lines typed so that these may be recalled for automatic correction, the printed character being lifted from the page.

The electronic typewriter can be provided with extra memory units to make it even more versatile. From here it is a small step up to the word processor.

The Word Processor

This is basically an electronic typewriter with memory connected to a videoscreen and printer. The operator types on the conventional keyboard in the usual way, and the words appear on the screen. There they can be repositioned, edited, replaced as required. When the text is perfect it can be printed out, perfectly.

For sending out circulars, for example, the same letter can be printed but with different names and address each time, these being taken from the word processor's memory on rigid or floppy discs. For increased memory capacity several word processors, at so-called work stations, can be connected to a central unit which has a huge memory. It acts as a central electronic filing system. Here word processor and computer merge. In fact, word-processing programmes are now available for many home computers.

In the latest machines the actual printing of the text is done by means of tiny jets of ink. Ink-jet printing is much faster and much quieter than the original mechanical impact method (like a typewriter). For high-volume printing, electronic printers may be used which work by means of lasers. They can print out information from word processors and computers on the spot or via the telephone at a distance. They can produce high quality images in any of hundreds of different typefaces.

Printing 2

Left: High-speed data printers form an essential link in modern business communications. Connected with computers, they are used both to transmit and receive information. These models can print at a rate of up to 150 characters a second.

Right: Colour printing is done using four plates, which print respectively yellow, magenta, cyan and black. The top three pictures show the separate colours applied by the three colour plates. The bottom picture shows the final result when the three colours and black are printed together.

Typesetting

In the traditional method of typesetting the words, or "copy" are set by assembling individual type characters (monotype) or lines (linotype). But these methods are fast becoming obsolete. Their place has been taken by film-setting, or photosetting. The type images are formed photographically.

In the latest photosetting machines, which are computer controlled, the operator taps in the words to be set on a standard keyboard and sees them displayed on a video-screen. He can then make any alterations before he commits them to a disc, where they are stored as magnetic "signals". This is then linked with a film unit, in which the signals produce the type image, usually on sensitized ("bromide") paper. The words stored on the disc can be "recalled" onto the keyboard by the operator at any time and altered as necessary. This makes the updating of information very easy. Laser typesetting machines are also coming into use.

Colour Separation

Lasers also feature in the latest electronic scanning machines used for colour separation work. Colour separation is the process by which coloured illustrations are processed for printing. It depends on the principle that any colour can be produced from combinations of the so-called primary colours red, blue, green and black. In the usual process the colours are separated by means of colour filters and screens. The electronic scanner analyses the colour by computer and produces the colour separations on film by means of a laser beam. It produces separations faster and more cheaply than other methods.

Printing plates are then made from the colour separations representing the amounts of red, blue and green in the original illustration. In the printing process the plates are coated with ink of the complementary colours, respectively cyan, yellow and magenta. When these inks combine, with black, they reproduce the original colours.

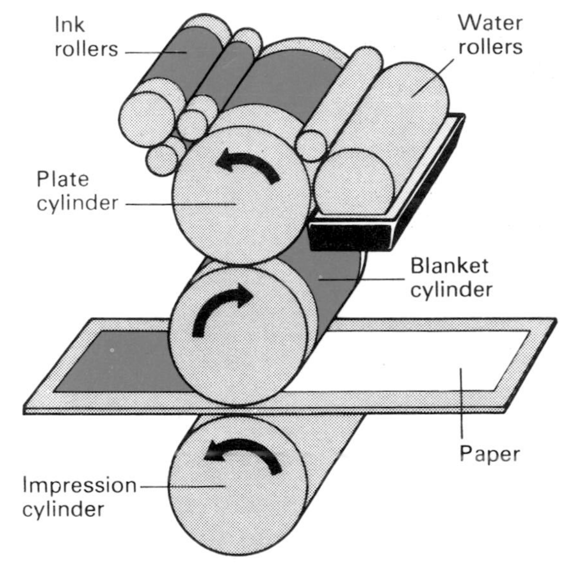

Above: An outline of the offset-litho printing process. The litho plate is wrapped around the printing cylinder and first wetted, then inked. The blanket cylinder then transfers the ink image to the paper.

Below: The video display unit of an electronic typesetting machine, in which the type characters are stored in digital form and generated by computer.

Printing Methods

The method Gutenberg used for printing is becoming obsolete today. Type characters are cast in moulds in a lead alloy called typemetal. They are coated in ink, paper is pressed against them, and the inked image is transferred to the paper. This method is called letterpress, and printing takes place from a raised surface. In another printing method, called gravure, printing takes place from a recessed surface. The type characters are cut into the printing plate. Ink is applied to the plate and a knife blade is scraped across it afterwards. The ink is retained only in the recesses, and is transferred when paper is pressed against the plate.

The commonest method in use today, however, is offset-litho, which uses a flat printing plate. The type image is transferred photographically to the plate in such a way that only where the type image falls does the plate attract the greasy printing ink. On the printing machine the plate is clamped around a cylinder and then wetted and inked in turn. The printing areas repel the water but attract the ink. The inked image is then transferred to another cylinder ("blanket"), which in turn transfers the image to the printing paper.

"Electronic" Publishing

The electronic revolution has spread even further in some areas of printing and publishing. In some newspaper offices, for example, reporters and editors work at video input units which are linked through a computer to phototypesetting machines and in some cases to automated platemaking machines.

The reporters use a word processor to type their stories onto the videoscreen and place them in the computer's memory. An editor can recall the story on his own screen and make any changes. He can then send it for setting in a photosetting machine. He may well be able to make-up, or design each page on the videoscreen before sending the whole unit for setting.

There are many other electronic variations to the system. Reporters once used to telephone their stories by word of mouth into the news desk where someone would take it down in shorthand and then type it up, hopefully correctly. These days many reporters type their stories on electronic typewriters, which feed them via telephone lines directly into the memory of the newspaper computer.

Also, completed newspapers and magazines may be transmitted in their entirety from one location to another by electronic means. At the receiver end, the signals are converted automatically into duplicate film, from which the printing plates are made. This system allows international magazines to be printed simultaneously in different parts of the world and greatly simplifies distribution.

Photography 1

Photography – "drawing with light" – has for many years been not only one of the world's most popular hobbies, but also an essential means of communication. How dull newspapers, books and magazines would be without photographs. Photography has also become a vital scientific tool. For example, it can record in invisible light details that the eye cannot see; it can "freeze" the motion of a humming bird's wings or a bullet in flight.

Photography is also vital in many other different fields – in medicine for taking X-ray pictures, in printing for making litho printing plates; in business for PHOTOCOPYING and microfilming; in the semiconductor industry for making silicon chips. It also makes possible the art and craft of cinematography, or MOTION PICTURES.

Above: The film disc of a Kodak disc camera. The film has been developed and the pictures appear as negative images.

Top right: The Kodak disc camera in use.

Right: Sony's Mavica electronic camera, which takes "pictures" on a magnetic disc. This is "played back" through a television receiver, and images appear on the screen.

Bottom right: A piece of microfilm. In microfilming, pages of information are greatly reduced photographically and stored as images on film.

Below: A picture taken by a fish-eye lens, showing the distortion that occurs.

The Film

In about 1826 a Frenchman, Joseph Nicéphore Niépce, produced a crude photograph on a pewter plate coated with bitumen. But this was not a practical method of photography and attention focused instead on methods using silver salts. Silver salts, particularly silver iodide and silver bromide, still provide the chemical basis of photography today. They are the essential ingredients in the coating, or emulsion on photographic film. When they are exposed to light, they are altered chemically by it and retain an invisible image of what they "saw".

During the process of photographic developing, this image is made visible. The light-altered silver salts are changed into metallic silver on the film, and this forms the "negative". By shining light through the negative onto photographic paper and developing the paper, a "positive" is formed which is an exact representation of the scene "viewed" by the silver salts.

Colour film contains three layers of emulsions, which record respectively the amount of blue, green and red in the scene viewed. In developing and printing the layers are dyed in suitable colours (yellow, magenta and cyan), which combine as far as the eye is concerned to reproduce the original colours viewed.

The Camera

This is essentially a light-tight box with the film on one side and an opening (aperture) on the other. There is a lens in the opening to focus the light rays onto the film, and a shutter behind the lens which opens for a fraction of a second to admit the light. The simplest cameras to use have few if any controls to alter the exposure, or amount of light reaching the film. Likewise, they have no means of focusing the image on the film. However, they have a very small lens aperture, which allows them to focus on everything beyond about a metre or so away. Other simple-to-use cameras have one or two exposure controls or maybe an electronic eye, which automatically sets the exposure according to the amount of light available. For dim lighting conditions the camera may incorporate an ELECTRONIC FLASH unit.

One of the latest "fool-proof" cameras is the Kodak disc camera. Whereas most other simple modern cameras use a cartridge film pack, the disc camera uses film on a revolving disc. After each picture has been taken a motor turns the disc ready for the next exposure. Exposure is set automatically and the built-in flash unit is activated if the lighting is dim.

Reflex Cameras

For best results, however, photographers need to be able to alter the exposure of the film, to allow for different lighting conditions; to vary the speed of the shutter, to prevent blurring when something in the picture is moving; and to focus, to make the picture sharp. All advanced cameras have means of doing all three. They have controls to vary the aperture and the shutter speed, and a lens that can be moved in and out for focusing.

The commonest type of camera with variable controls is the single-lens reflex (SLR). This is so called because it uses a mirror to reflect light coming through its one lens onto a focusing screen. When the shutter release button is pressed, the mirror "flips up" out of sight, allowing light to fall on the film.

The most popular SLR is the 35-millimetre (mm) camera which takes a cassette of film 35 mm wide holding up to 36 exposures. It has typically about nine aperture settings and about the same number of shutter speed settings down to one-thousandth of a second. Larger format SLRs like the Hasselblad, use 120 roll film and take pictures about 55 mm square. This film is also used in the twin-lens reflex cameras, which have a separate viewing lens.

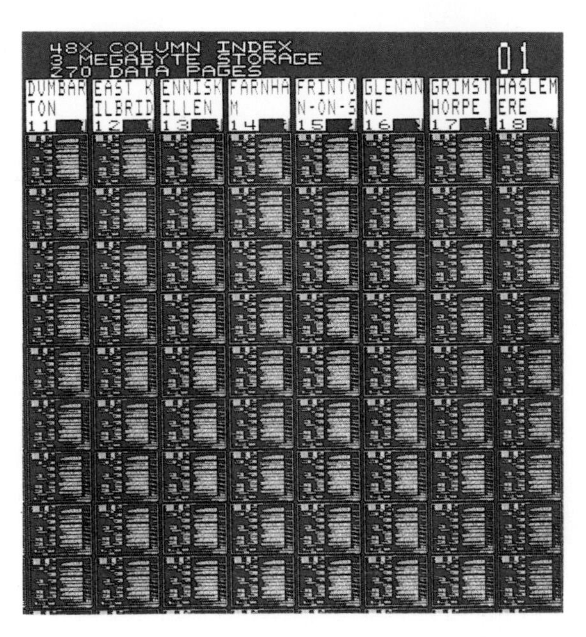

For the best-selling SLR cameras like Canon, Nikon and Olympus a vast range of interchangeable lenses is available. Telephoto lenses allow the photographer to photograph at long range; wide-angle lenses allow the capturing of a broader field of view; zoom lenses allow a change of focus from short to long range.

The modern SLR cameras also incorporate an accurate exposure meter, which is coupled to the camera aperture and shutter-speed controls. In some models the exposure is set automatically by the meter via electronic circuits. In others the photographer alters either aperture or speed settings until the exposure indicator shows that these settings will give the best exposure. The indicator may be a needle moving over a scale or a series of light-emitting diodes (LED), of the type often used for the display in some pocket calculators. The photographer can read the indicators from inside the camera viewfinder.

Photography 2

Instant Picture Cameras

Developing and printing ordinary film can take several days through the shops. This is what prompted the inventor Edwin Land in 1948 to develop the Polaroid "instant-picture" camera. This type of camera has been progressively developed since then into an ingenious picture-taking and processing device known as the SX-70. The SX-70 is a kind of folding reflex camera with automatic exposure control, and in the latest model, automatic focusing.

The Autofocus model uses sonar for focusing. When the shutter button is pressed, an ultrasonic beam is transmitted from the camera. It is reflected by the object to be photographed back to the camera. The time between transmission and return is measured by a built-in clock. From this time of travel, the distance to the object is calculated and the camera lens is automatically moved in or out as necessary. Then the shutter is opened to expose the film. All this takes place in a fraction of a second.

The SX-70 is loaded with a film pack for 10 exposures. The film consists of many layers of emulsion and dyes sandwiched between plastic sheets. It also incorporates a small pod of developing chemicals. After a picture has been taken, a piece of film is ejected from the camera through a pair of rollers. The pod of chemicals bursts and starts processing the film. In about a minute for a colour film, processing is complete, having taken place outside

Above: An aerial photograph of farm crops in Indiana, USA, taken on film sensitive to infrared light. The vegetation shows up as shades of red. Differences in crop colour could indicate the presence of disease.

Viewfinder

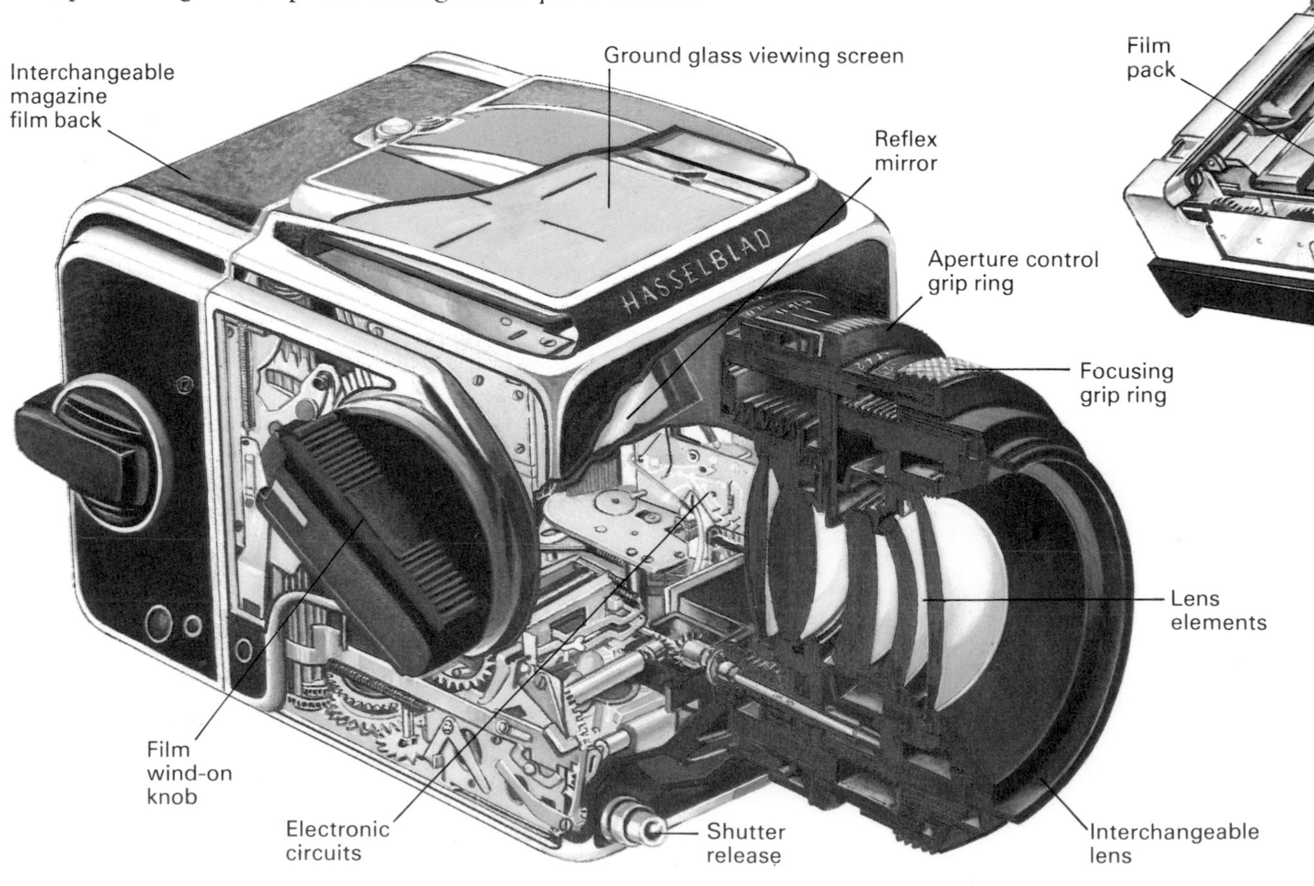

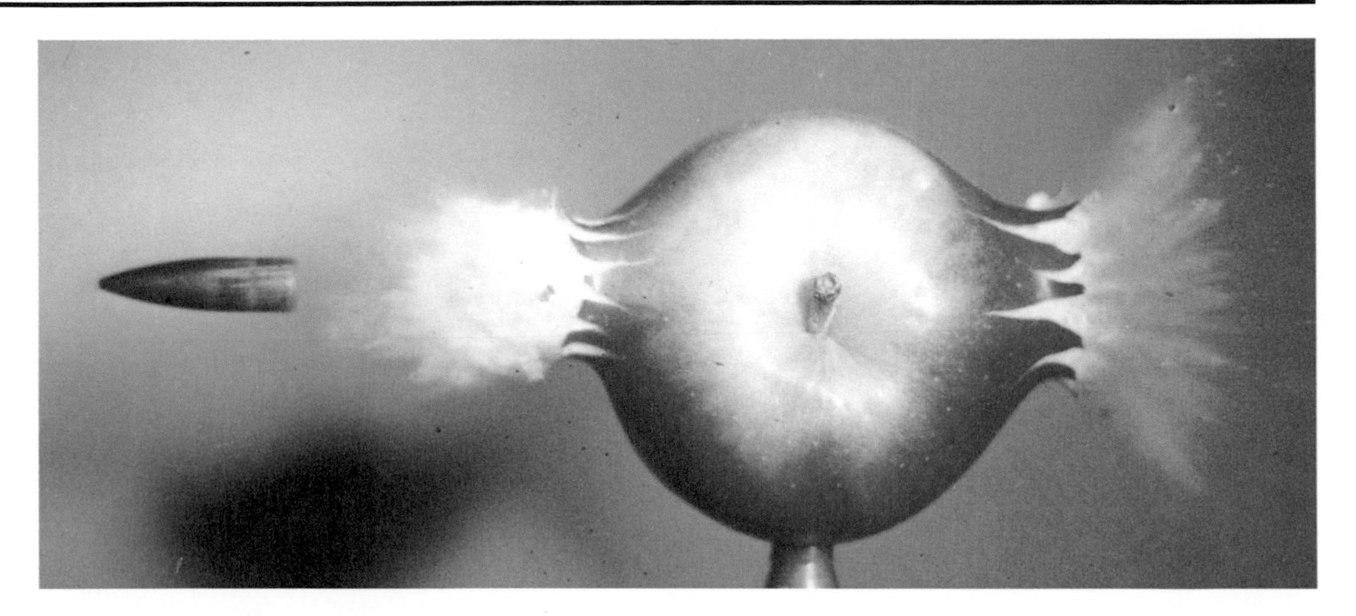

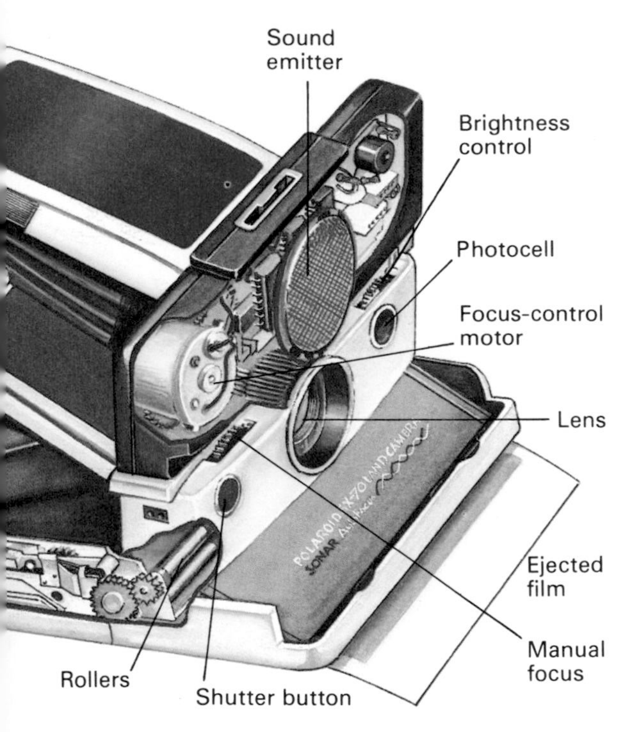

Above: A cutaway view of the Polaroid SX-70, which focuses by means of sonar.

Top: High speed photography can "freeze" the motion of fast-moving objects like this bullet. The results can be very spectacular.

Left: The 2000FC model Hasselblad, one of the finest of all cameras. It is a single-lens reflex camera, which carries roll film in an interchangeable back. This model has an electronic "brain" and has shutter speed settings down to 1/2000th of a second.

Right: An ultra-long telephoto lens like this gives greatly increased magnification, but only a narrow field of view. To prevent camera shake, the equipment must be sturdily mounted on a tripod.

the camera in the light. This is a far cry from the traditional way of processing films, which must be done initially in the dark and requires a lengthy sequence of operations.

Electronic Cameras

One of the latest innovations in home photography has been the introduction of electronic, or filmless cameras such as Sony's Mavica. The Mavica looks much like a normal 35-mm camera, but it records the image coming through the camera lens on a spinning magnetic disc. To "print" the picture taken, the disc is inserted into a device that plugs into a television set. The "photograph" then appears on the screen.

Another type of electronic camera is used in ultra-high speed photography for scientific work. It is called an image converter camera. It is able to make exposures of as little as one-million-millionth of a second. This means that it can "freeze" the movement of the fastest moving objects. The camera works much like a television camera. The light entering the camera lens falls on a plate, which emits electrons where the light falls. The electrons then form an image on a television-type screen which is then photographed normally.

Audio, Video 1

Many advances in science and technology are not immediately obvious to the ordinary man in the street. But those in the field of electronics certainly are, for we see the results in our own homes: in pocket radios, colour television, hi-fis, video games, video recorders, home computers, and so on. These devices belong to a branch of communications we often call audiovisual, or audio (sound) and video (picture).

The Radio

The radio was the first means of mass communication, which grew up in the early years of this century. In radio broadcasting sounds are converted by a MICROPHONE into electrical signals. These "audio" signals are then transmitted through the air on a radio "carrier" wave. The aerial of a radio receiver picks up the carrier wave, and its electronic circuits remove ("detect") the audio signals and feed them to a LOUDSPEAKER, which changes them into sound.

Radio broadcasting uses a variety of radio waves of different wavelength or frequency, which allow different programmes to be broadcast at the same time. Low-frequency waves are used in general broadcasting. They can travel long distances around the curve of the Earth because they are reflected by layers in the atmosphere. Short, medium and long-wave radio broadcasts use wavelengths from about ten to several thousand metres.

For highest-quality broadcasting very high frequencies (VHF) radio waves are used. But they have limited range because they cannot travel over the horizon. VHF bands are also used for most two-way radio communications, for example, in taxis, mobile patrol cars, the armed forces, and citizen-band (CB) radios. Radio frequencies are expressed in hertz (cycles per second). VHF broadcasts take place typically at about 80–90 million hertz (MHz).

The Television

VHF and UHF (ultra-high frequency) bands are used for television broadcasting. The technique is broadly similar to radio broadcasting: electrical signals representing the pictures to be transmitted ("video" signals) are carried through the air on a radio carrier wave. The aerial of a television receiver picks them up, and electronic circuits separate out the video signals, which are then fed to the picture tube. There they cause an image to be formed on the screen.

The video signals are produced by a television camera. The lens of the camera forms an optical image of the scene viewed, which is then turned into an electrical image. A beam of electrons from an electron gun scans the image from left to right and from top to bottom in several hundred lines (usually 625 lines in Europe, 525 in the United States). The result of scanning is a variable electric

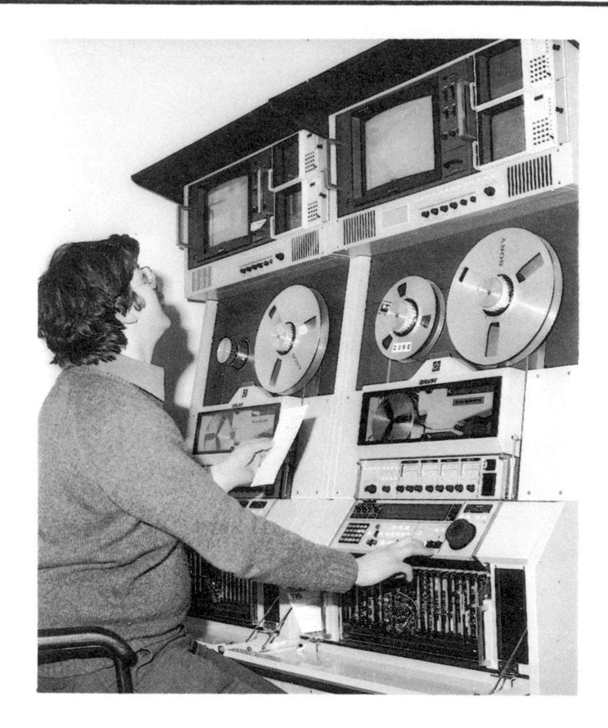

Above: A video tape recording (VTR) console used to record programmes in a television studio. The video recorders use wide magnetic tape, which is threaded from one reel to the other through the recording head.

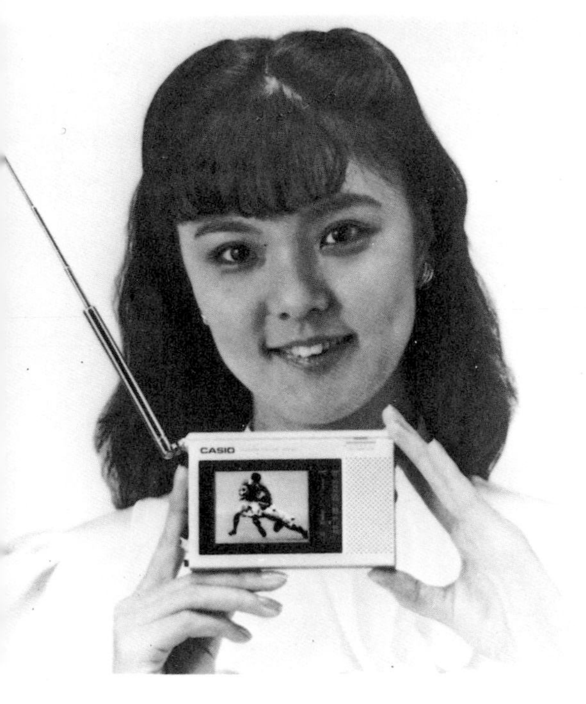

Above: A pocket-size flat-screen TV. This model uses a liquid-crystal display, similar to that in most digital watches and calculators.

Below: Home-hunting by Prestel, Britain's view-data service. Viewdata users can call up thousands of "pages" of information onto their TV screens using a hand-held keyboard.

current that represents the light pattern viewed by the camera. These are the video signals that are transmitted.

The picture tube in the receiver is a form of CATHODE-RAY TUBE. In the tube the video signals are made to alter the strength of an electron beam from a gun, scanning across the fluorescent screen. The result is a pattern of light on the screen similar to that viewed originally by the camera. The ordinary picture tube is large and bulbous, but recent advances in technology have resulted in much flatter tubes becoming available.

For colour television, three sets of signals are transmitted and received, representing the amounts of red, blue and green light in the various parts of the scene viewed. In the picture tube three guns are used to form red, blue and green coloured dots on the screen. These merge into the correct colour when viewed from a distance.

In most countries there is a limited number of television channels, but in the United States there are many, both local and national. Some are received through the air in the normal way, but the others are channelled through cables. Cable television will be a major growth area in other countries in the near future. So will satellite television – the direct broadcasting of television programmes from a satellite into the home. This is becoming possible because of the greater power of the communications satellites now being launched.

Videotext

The ordinary home television set can also be adapted to act as a data terminal to receive all kinds of information – the latest news and weather, current foreign exchange rates, the menus in local restaurants, road-traffic information, airline schedules, and so on. There are two basic systems for doing this, generally called teletext and viewdata. They are sometimes considered together under the heading videotext.

Teletext signals are transmitted with the ordinary broadcast television signals. They appear in code in a few spare lines on the television screen which are not used for the picture. Viewers can call up teletext information onto the screen by pressing a button on a remote-control device. The information is presented in the form of numbered pages. The viewer selects the page number he requires from an index and keys this number on his control device.

The BBC pioneered the use of teletext in 1976 with their system called Ceefax. Britain's Independent Broadcasting Authority launched a similar system called Oracle two years later. Teletext systems are now in operation in several countries. The American teletext system is called Infotext; the Canadian system, Telidon; and the French system, Antiope.

Audio, Video 2

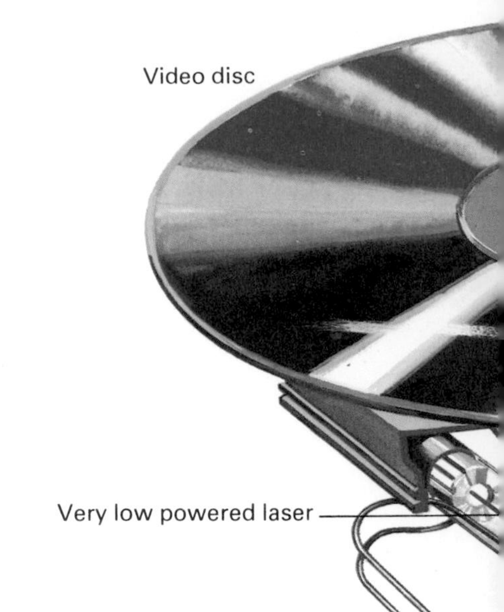

Viewdata

Teletext systems only offer a few hundred "pages" of information, however. Much more comprehensive information can be obtained via viewdata. In this system, the information is displayed on the television set but is carried by the telephone lines. The viewdata user does not have to dial any numbers, however, but calls up the service with a remote-control keyboard. He and his television are then hooked up to the viewdata computer which has probably hundreds of thousands of pages of information stored in its memory. The viewdata system was again pioneered in Britain by British Telecom, whose present system, Prestel, is still a world leader.

Audio Recording

The other side of audio-video – recording – has also undergone quite a revolution in recent years. In audio, or sound recording, the quality of reproduction from both disc and magnetic tape can now be outstanding, or of very high fidelity, thanks to advances in electronics.

A typical high-fidelity (hi-fi) system is made up of a number of electronically matched units, comprising a record player, a tape deck, a tuner (to receive radio broadcasts) and an amplifier. The amplifier feeds signals from the other units to a pair of loudspeakers which are positioned to the left and right of the listener. This STEREOPHONIC arrangement produces "left" and "right" sounds similar to those that the listener would hear if he were listening to the sounds "live". Some hi-fi systems are wired for quadrophony, with an extra pair of speakers to be positioned behind the listener.

The tape deck now takes compact tape cassettes rather than large reels of tape, as in the old machines. Tape cassettes have begun to take over from discs as the most popular form of sound recording. They are easier to

Top left: Radio and closed-circuit television (CCTV) being used on a heavy-lifting vessel in the North Sea oilfields. The vessel is using the CCTV to help manoeuvre during the placing of a production rig.

Top right: The principle of the video disc system. The beam from a very low power laser passes through a series of lenses and mirrors and is focused onto the disc. It is reflected by pits in the disc, and the reflected beam passes to a photocell. This cell produces signals which are fed to the television receiver. A similar system is used to play the new type of compact audio record discs.

Below: A home video camera, which uses a series of filters to produce the video colour signals.

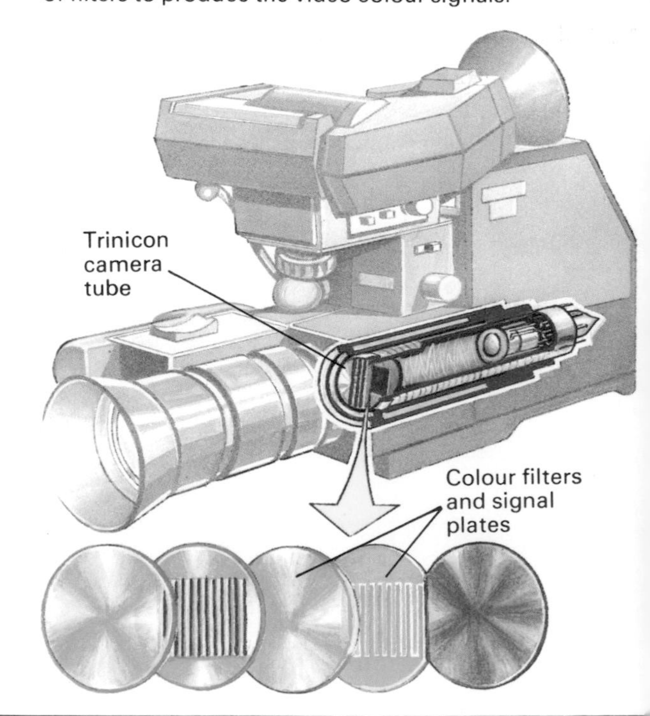

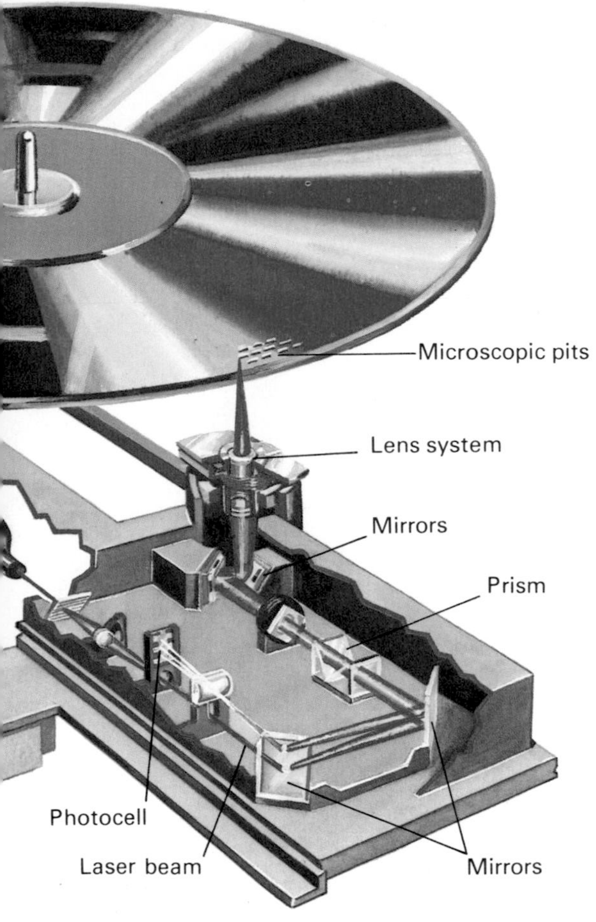

handle and less liable to be broken and damaged than discs and can be played in easily portable cassette recorders.

But for quality of reproduction, the disc has the edge, especially the latest type of disc, the digital disc. It is so called because sounds from the recording microphone are converted into signals that are represented precisely in terms of figures, or digits. And these precise digits are then used to cut the record.

Video Recording

Since the early 1970s tape cassette recorders have also extended the range of home television viewing. The video cassettes are much bigger than the audio cassettes because they have to include much more information, picture as well as sound. This is recorded as a zig-zag pattern on the 13-mm wide tape. You can record any channel you like with the video tape recorder, whether you are watching it or not; you can set the recorder to record programmes while you are out, and play them back when you want. You can also buy or hire pre-recorded tapes of favourite films. If you have your own video camera, you can even make your own television films.

As in many other advanced electronic fields, the laser is also making its mark in video recording – in the video disc. The video disc looks at first like a conventional disc, but is smooth to the touch and has no grooves. Picture and sound signals are recorded in the form of microscopic pits in a thin layer of aluminium beneath a plastic coating.

The video disc is not played with a stylus (needle) like an audio disc, but is scanned by a laser beam. The beam picks up the sound and picture signals as it is reflected from the pits.

Bottom right: A home video camera in use.

Below: The essential parts of a Trinitron television receiver. Unlike most colour TVs, it uses a single electron gun, which produces three beams. The vertical grille and colour phosphor stripes are also different from other systems.

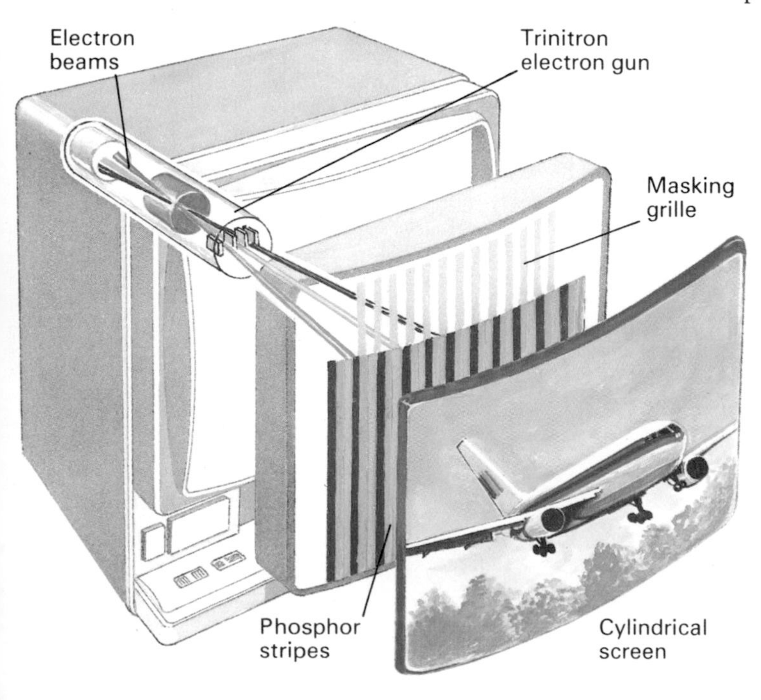

Telecommunications 1

One notable feature of our age is "instant" communications. We can pick up a telephone, dial a number, and speak to someone in another town or thousands of kilometres away in another country, or another continent. We can switch on the television and see events happening "live" virtually anywhere in the world. In busy offices Telex machines send messages to, and receive them from offices scattered far and wide. Other machines transmit and receive all kinds of other data (information) – weather reports, computer print-outs, banking transactions, FACSIMILE documents, and so on.

The field of long-distance communications is known as telecommunications ("tele" means "far"). American inventor Samuel Morse started the telecommunications revolution in the 1830s when he invented the electric telegraph. He sent coded electrical signals down a wire, which could be received and interpreted as a message. The modern equivalent of the telegraph is Telex. A machine called a teletypewriter, or teleprinter, is used to code a message which is then transmitted. It is received on a similar machine, which decodes the message and types it out automatically.

Some 40 years after Morse, Alexander Graham Bell invented the telephone, making it possible to send voices down a wire. In the telephone the voice is converted by a microphone in the mouthpiece into electrical signals which are transmitted to the receiver (earpiece) of the person being called. There they are converted into sound waves again.

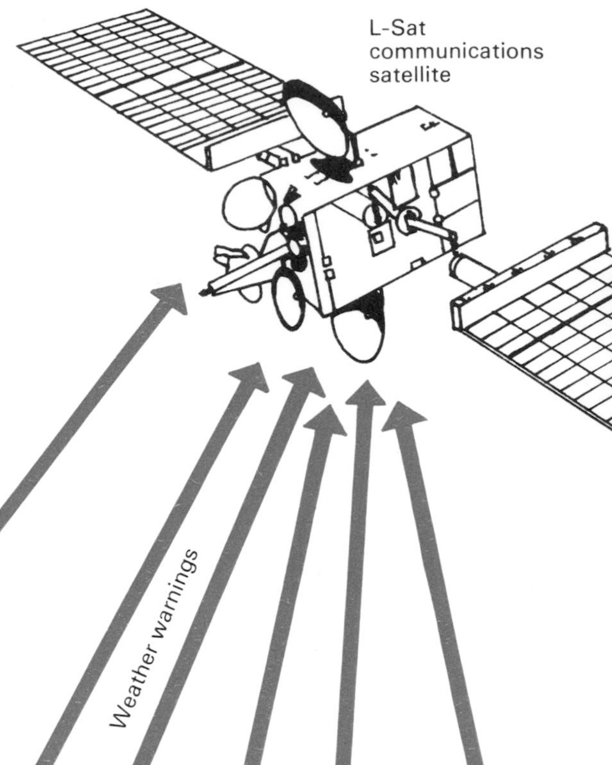

Satellites high above the Earth are playing an ever-increasing part in communications of all kinds. Through Intelsat (the international telecommunications satellite organization), they handle television, telex and telephone signals. Through Inmarsat (the international maritime satellite organization), they handle routine and emergency communications at sea.

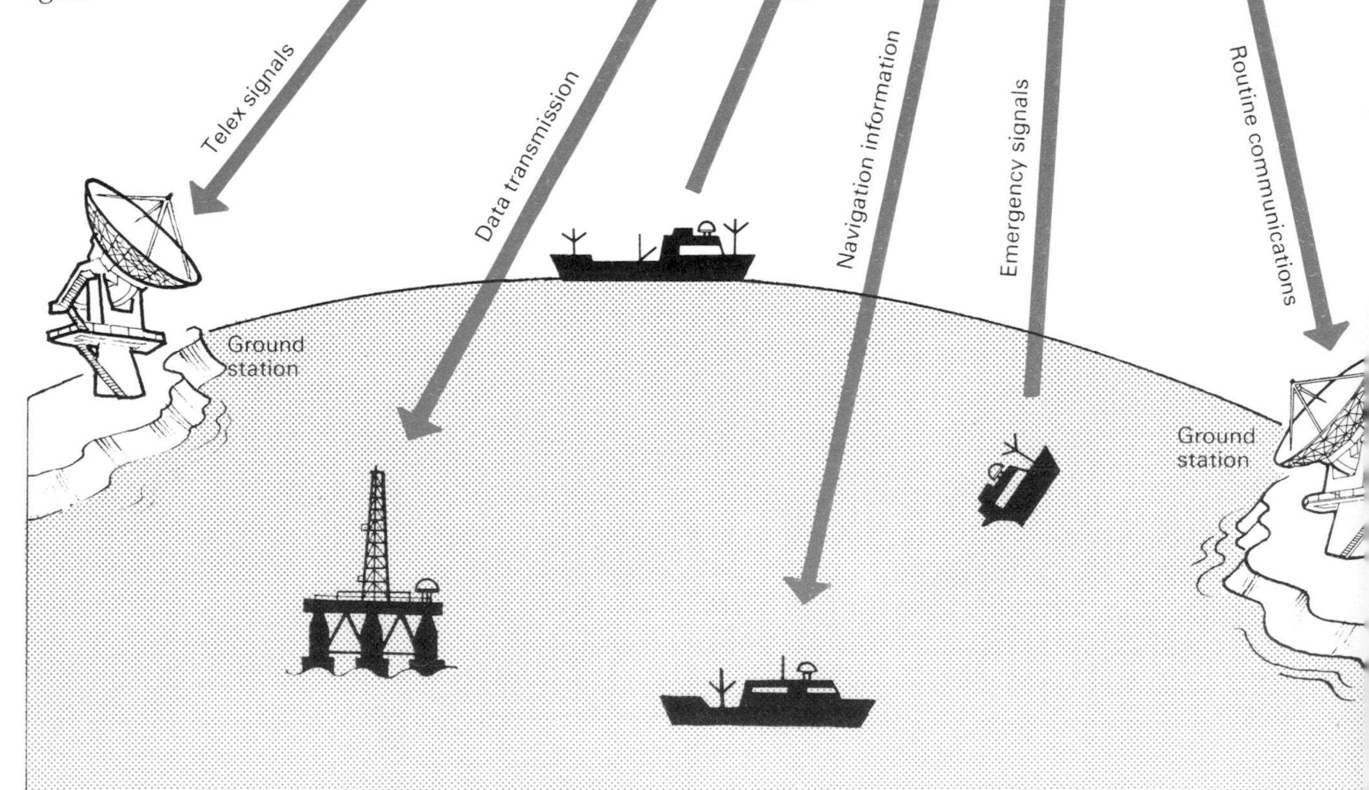

Above: Two photographs that illustrate the old and the new types of telephone exchanges. On the left an engineer is testing the circuits in the latest type of computer-controlled exchange, known as System X. Connections are made by means of electronic switches. Early this century (*right*) telephone operators made connections manually on switchboards like this.

Space-age Techniques

Twenty years ago most long-distance communications by telephone and telegraph went by way of copper cables overland, underground and undersea. People wishing to make overseas calls had to book them in advance or wait until there were lines free because the submarine cables had limited capacity. And when they made their call, it was often of very poor quality.

Today, however, we can usually dial international calls ourselves, and the quality of reception is generally excellent. The calls may still go by cable for part of the way, but they will probably travel for the most part by MICROWAVES (very short radio waves), via space. They first go to an international exchange, from where they travel by cable to a microwave relay tower. They are converted to microwaves which are then beamed by way of other relay towers to a satellite ground station.

This beams the signals up to a communications satellite, such as Intelsat V, located 36,000 km high above the equator. The satellite strengthens the signals and beams them down to a receiving ground station, from where they travel, via more relay towers, to another exchange. This routes them to the receiver's telephone.

The telephone handset itself has been transformed in recent years. Telephones with push-button dialling are now commonplace. Cordless telephones, which are portable radio-telephones, are coming into use. Phones with silicon-chip technology may hold a memory for remembering frequently-dialled numbers, or show the number of the person calling. Some can be converted into computers.

Below: The technology of cabling is also undergoing a revolution as optical fibres replace copper wires. In this picture the thin fibre cable on the left can handle the same number of telephone calls as the thick copper cable on the right.

Telecommunications 2

Light Cables

The great expansion of telecommunications networks has caused problems with cabling. Modern cables are coaxial – they are made up of several cables in one. They can carry thousands of telephone calls at once, using a technique called multiplexing. But these cables are bulky and in many areas already fill the cable ducts built to carry them beneath the streets. And they are made of copper, which is becoming increasingly expensive and in short supply.

This has led telephone engineers to turn to two new technologies, FIBRE-OPTICS and lasers. They have designed thin threads, or fibres, of glass that can carry laser-coded messages long distances.

An optical cable containing a few fine glass fibres has the capacity of a bulky coaxial cable many times its size. An optical fibre one-tenth of a millimetre in diameter can carry 2000 two-way telephone conversations at once! Optical cables are already widely in use in North America, particularly for carrying cable television channels. One of the biggest systems is being installed in Saskatchewan, in Canada, where a 3000 km optical cable network will eventually connect all the major cities.

At the transmitter end of an optical-cable network, telephone signals are converted into pulses of pure laser light. These pulses pass through the glass fibres by reflection on the inside; virtually no light escapes from the sides. At the receiver end the laser pulses are converted back into electrical signals that pass to the telephone.

Above: A view of Mission Control, Houston, during the first Moon-landing mission in 1969. This was a triumph in long-distance communications. Several hundred million TV viewers, 385,000 km away on Earth, saw astronaut Neil Armstrong plant the first human footprint in the lunar soil.

Opposite: A spectacular array of antennae at the Carnarvon ground station in Western Australia. This facility forms part of the satellite communications and tracking network of the European Space Agency (ESA).

Left: Optical fibres being used to carry television signals. The signals are transmitted in the form of coded pulses of laser light.

Navigation

Above: A radar installation in the Shetlands. It was set up to provide control of air traffic in the North Sea oilfields.

Below: High-definition radar from space can provide clear images of the Earth's surface at night and despite cloud cover. This picture shows a radar image, at night, of Lake Okeechobee, in Florida.

It is relatively easy finding your way about the countryside where there are recognizable landmarks. A ship's captain sailing in the middle of the ocean or an airline pilot flying high above the clouds are not so fortunate. They have no landmarks to guide them.

However, thanks to modern science, they still know where they are. Many methods of finding the way, or navigating, have changed little with the years. The compass is still widely used; so is the sextant. But increasingly, faster and more accurate electronic navigational methods are taking over.

The compass is a useful aid to navigation, because its needle always points in the same direction, north–south. These days, however, the magnetic compass has been largely replaced by the gyrocompass, which contains a rapidly spinning gyroscope. It uses the property of the gyroscope that, once set spinning, it always stays pointing in the same direction no matter how its supporting frame twists and turns.

Inertial Guidance

Gyroscopes are also essential parts of the inertial guidance system by which submerged submarines, guided missiles and spacecraft navigate. This system uses three spinning gyroscopes, three accelerometers and a computer. The gyroscopes provide a reference "grid" in the three dimensions. Devices sense the movements of the submarine (say) in relation to this grid and feed the information to a computer.

The accelerometers continually sense the submarine's acceleration, that is, changes of speed, in the three dimensions, and also feed this information to the computer. This

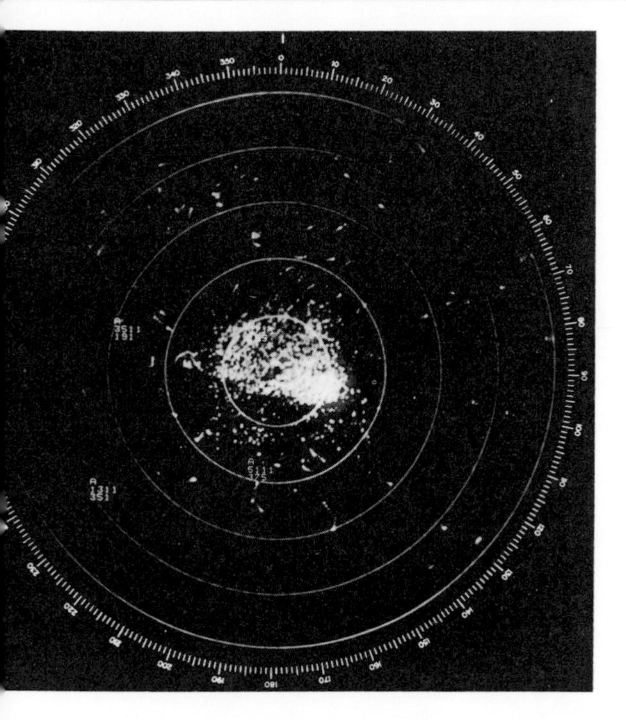

Above: The screen of an air-traffic control radar unit. The aircraft are identified by number. In practice this information is presented in three colours.

Below: An Intersub submersible craft of a type now widely used in offshore oilfields. Submersibles rely on sonar for both navigation and communications.

now has all the information necessary to work out the exact position of the vessel at any time. The system can be set to work automatically, by instructing the computer to remain on a pre-set course. This works rather like the AUTOMATIC PILOT of an aircraft.

Electronic Aids

One of the first electronic aids was the radio direction finder (RDF), which is still widely used. In this system radio beacons at various locations send out signals. Each is coded differently for identification. On board a ship the bearings (directions) to two such beacons are determined by a RDF. From these bearings the position of the ship can be pinpointed.

More advanced RDF systems are Decca and Loran, which are used widely by ships and aircraft. In these systems ground beacons send out specially phased radio signals. Equipment on board ship or aircraft uses these signals as a reference grid and automatically displays coordinates, which pinpoint the craft's position to within a few metres.

Satellite navigation systems are also coming into use now for civilian as well as military purposes. Satellites such as the American Transit and Navstar send out position and timing signals which ships use for reference. Other satellites provide ships with up-to-the-minute navigation information through the communications network set up by the recently formed (1982) INMARSAT (International Maritime Satellite Organization).

Radar and Sonar

A ship's captain not only wants to know exactly where he is at any time, but also where other vessels are in order to avoid collisions. Unlike a car driver, he cannot apply the brakes and stop quickly. Large tankers take hundreds of metres to stop. So most vessels, large and small, are now equipped with navigational RADAR. Radar is widely used for many other purposes relating to navigation such as air-traffic control at airports. Planes often have weather radar, which detects the presence of storms that can then be avoided.

Radar equipment sends out pulses of very short radio waves (microwaves), which are reflected by any objects in their paths. The reflections, or "echoes", are received and displayed on a screen, which is similar to the screen of a television receiver. The position and direction of the echoes on this screen indicate the range and bearing of the object.

SONAR is a similar system used for navigation underwater, but it uses sound waves instead of radio waves. Surface ships use sonar in the form of an ECHO-SOUNDER to measure ("sound") the depth of water beneath the hull.

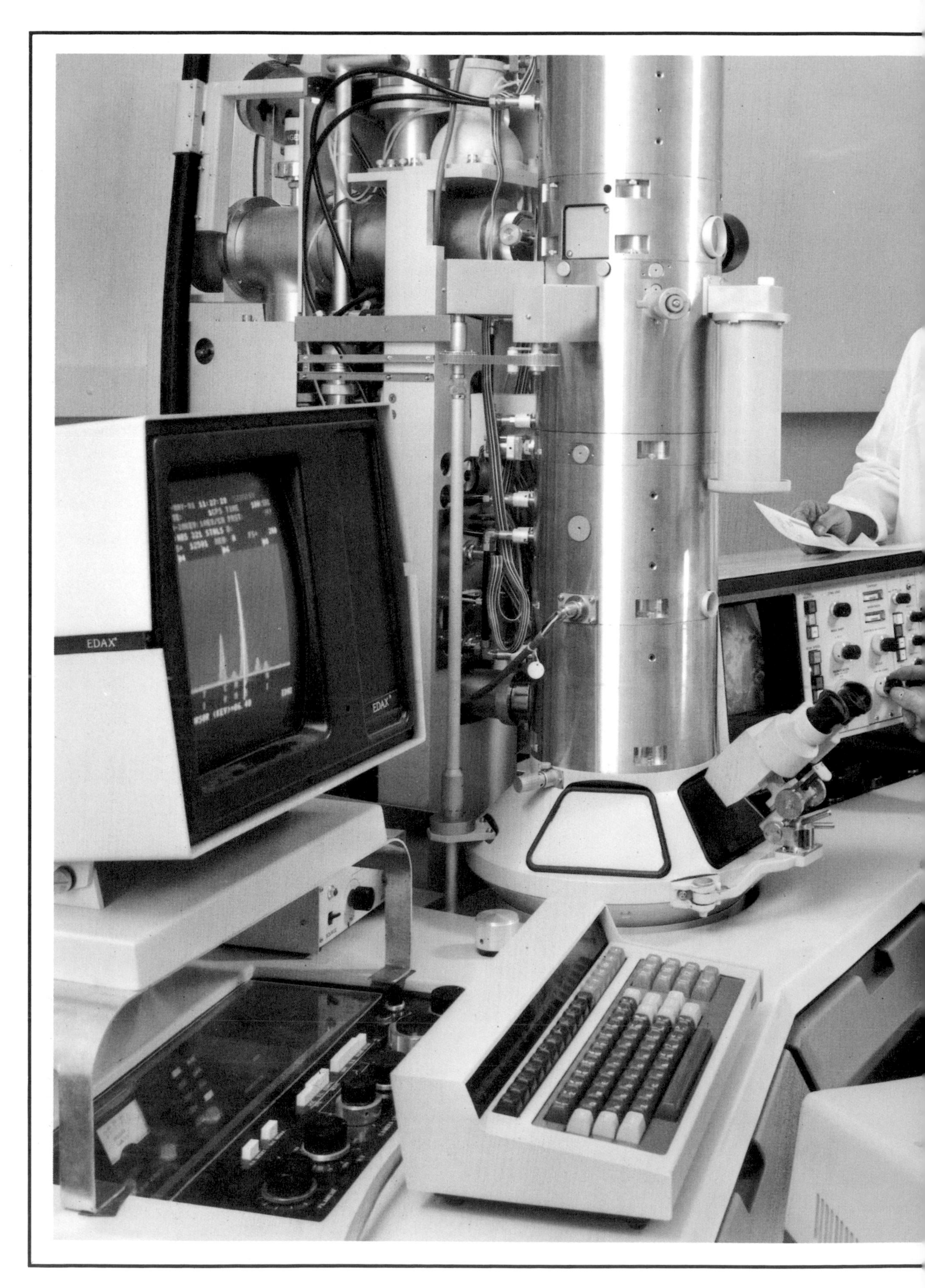
EDAX
EDAX

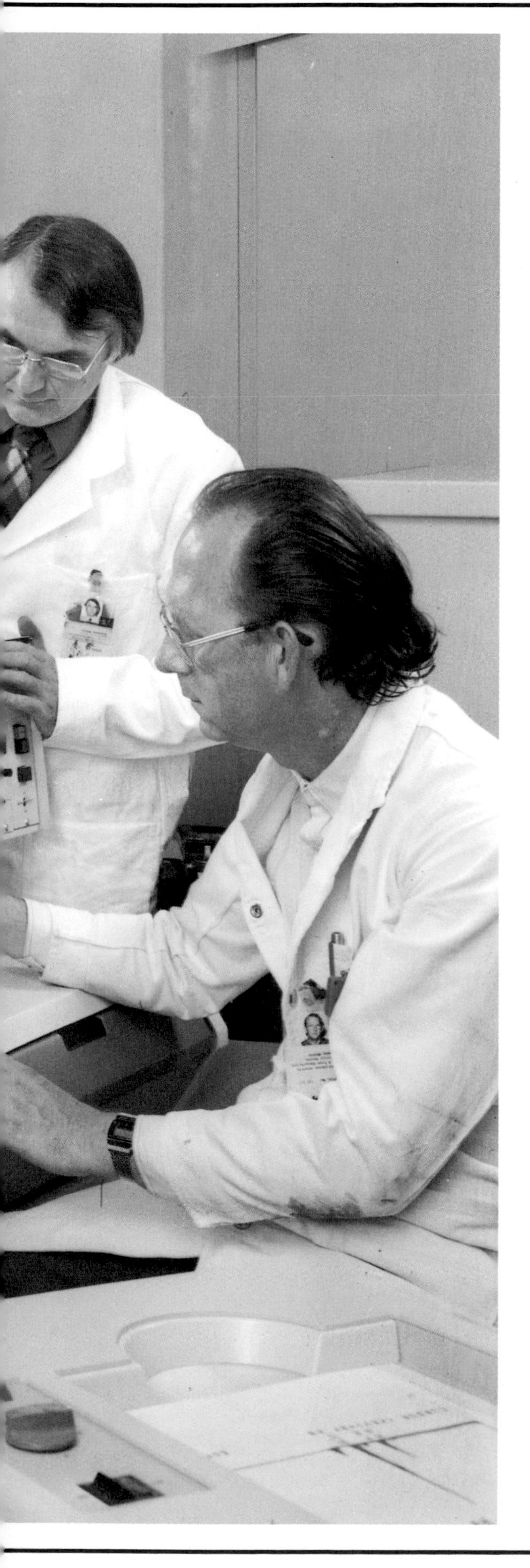

Chapter 8

Science and Medicine

The scientific discovery of today becomes the technological miracle of tomorrow – if all goes well. The discovery of the peculiar electrical properties of silicon in the 1940s led, via the transistor, to the wafer-thin silicon chip and the microprocessor. "Microprocessors and chips" are now part of the staple diet of science and technology.

Research into the atomic nature of matter earlier this century with atom smashers and other devices led to the discovery of nuclear fission. We now harness this process in nuclear power stations to produce electricity. The manipulation of atoms in rubies led to the development of the laser.

Now lasers have found a wide variety of applications in industry, engineering, telecommunications, video recording and medicine. Armed with ever more sophisticated instruments, scientists continue to blaze trails into the unknown, with the technologists hard on their heels.

Researchers working with a high-voltage electron microscope and a variety of display and data analysing equipment. This microscope is of the transmission type, examining specimens by passing electrons through them.

Silicon Chips

About 200 years ago the steam engine was the force behind the Industrial Revolution, which was to change not only industry but the whole of society. We are now at the beginning of another revolution that will also change our lives. This time it is not a powerful, steaming engine that is ringing the changes, but a thin wafer of crystal, small enough to pass through the eye of a needle. It is the silicon chip.

The chip is the "brains" behind the digital watch and pocket calculator. It has made possible the home computer and the fascinating variety of video games. It can play chess well enough to challenge a Grand Master; make a robot weld with inhuman precision; and guide a spacecraft to a rendezvous with a planet hundreds of millions of kilometres away.

Silicon chips can carry out such tasks because they function, either singly or in combination, as computers (see page 146). Some chips have only one function, serving as one part of a computer, such as the memory or central processing unit. The most advanced type of chip, however, by itself performs all the different functions of a computer. It is called the microprocessor, or microchip.

As with all computers, microchips can be "taught", or programmed, to do a variety of tasks. In particular, they can control the actions of other equipment. Because they are so small, they can be built into virtually anything to control it. In industry this is leading to increased automation. In the home it is leading to easier-to-use and more efficient-to-run equipment, from sewing machines to cars.

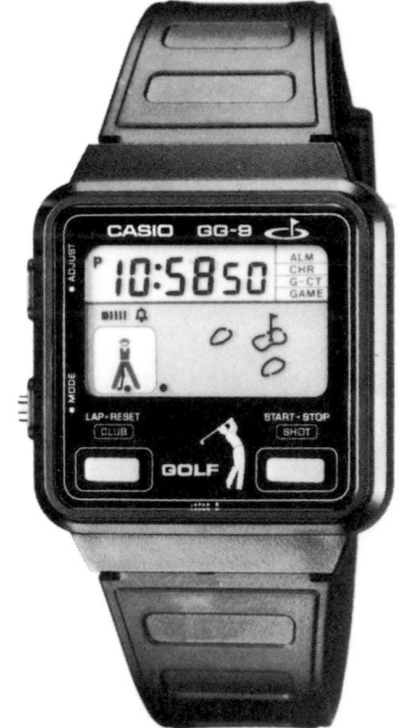

The Digital Watch

The digital watch is the most familiar object made possible by the microchip. It is a miniaturized version of the quartz clock (see page 157), which uses a thin wafer of quartz crystal as a time regulator. When stimulated by electric current from a battery, the wafer vibrates exactly 32,768 times a second. The electronic circuits on the chip then "count" the vibrations into signals that represent hours, minutes, seconds, and fractions of seconds.

In the digital watch the signals trigger off a display that shows the time in digits. Most digital watches now have a black-on-white liquid crystal display (LCD), which shows continuously. Many digital clocks and display devices have a red-on-black light-emitting diode (LED) display. In the alternative type of quartz watch – the quartz analogue watch – the time signals are made to drive hands, as in the ordinary mechanical watch.

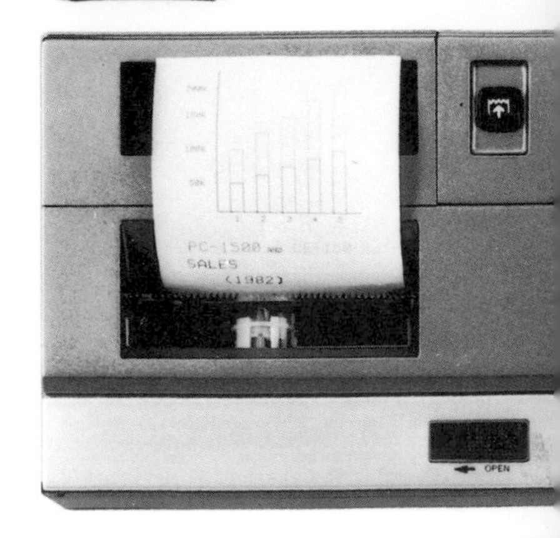

The Micro World

A typical chip is only about 7 millimetres square and weighs less than one-hundredth of a gram. Yet it contains

Above: A chip-making factory at Kosugi, in Japan. In this room the masks are being aligned on the silicon slices by microscope before they are doped. The premises are kept ultra-clean to prevent contamination of the chips.

Top left: A magnified picture of a chip held between thumb and finger. This is a 64K RAM chip, which means that it is used as a random access memory (RAM) for a computer and can store 64,000 "bytes" of information.

Left: Silicon-chip technology makes possible multifunction watches, which not only tell the time but also serve as stopwatch, calendar and alarm. On this model you can also play a 9-hole golf game.

Below: A silicon chip is the "brains" behind this pocket-sized microcomputer. This has an LCD display and is shown linked to a small four-colour printer.

thousands of miniaturized electronic circuits (microcircuits) and tens of thousands of electronic components, such as TRANSISTORS and RESISTORS. Some of the most advanced chips have nearly half a million components, and are capable of performing that number of calculations every second!

How is it possible to cram so many components into such a small area? It is possible by making all the components in the circuits from the same slice of material – silicon crystal. This process is called integration, and the circuits formed are called integrated circuits. The process of getting tens of thousands of components on tiny chips is called large-scale integration (LSI). It can take up to 2 years to design a LSI chip from scratch.

Silicon is one of the peculiar substances called SEMICONDUCTORS. It can be made to behave in a different way electrically by treating it with various chemicals. This process is called doping. When various combinations of the differently doped silicon are put together, they become different electronic components.

Chipmaking

The making of chips is quite a lengthy process. They are made, hundreds at a time, on a circular slice cut from a silicon rod. The circuits in the silicon are built up, layer by layer, in a sequence of 12 or more steps. Each layer is made by doping certain areas with chemical vapour. The other areas are covered by a photographic masking process.

In this process the silicon is coated with a light-sensitive layer, like photographic film. A mask is placed on top, and light is shone through it. Next the layer is developed and then etched with acid. Etching exposes areas of silicon for doping. This process of coating, masking, developing, etching and doping is repeated for each of the 12 or so layers. The masks used are first drawn some 250 times larger than life. Then they are reduced in size photographically.

When all the layers are completed, another masking and etching process takes place, using a metal such as aluminium or gold. This forms the connections between the various components of the circuits in the silicon. The chips are then inspected by a probe, and any defective ones are marked. The slice is next cut up into individual chips, which are mounted on a base. Fine gold wires are connected between the chips and pins on the base, which fit into a bigger circuit board.

The chip is now ready to play "space invaders" on the video unit; time the lap of a racing car; calculate cube roots and logarithms; set the exposure controls on a camera; record the purchases at a supermarket; correct mistakes on a word processor; or control a robot. You name it, a chip can do it.

Computers 1

Above: The main control centre for the Alaskan oilfields at Prudhoe Bay in the Arctic. A very powerful computer is in control and keeps a constant check on all the operations taking place. The human supervisor can call onto his video screen data relating to conditions in any part of the system.

Below: Computers are usually built up from electronic modules like this. They consist of silicon chips and other electronic components mounted on printed-circuit boards.

When the space shuttle thunders into orbit, it is not the astronauts who pilot it. When the trains on the new Lille Métro pull into a station, no human driver applies the brakes and opens the doors. When we receive a statement from the bank showing the cheques we have cashed, no human hand prepared it. What flies the shuttle, controls the Métro and prepares bank statements are computers.

Computers are now to be found in every walk of life. Multinational companies use them to calculate the payroll of millions of employees, and record and analyse their production and sales figures and other information. Meteorologists use them to make more accurate weather forecasts. Engineers use them to help design aircraft, cars and bridges. Supermarkets use them to prepare the check-out bill automatically and keep control of goods in stock.

However, computers are no longer just tools for industry and commerce. They are now available at low cost for use at home. The sales of home, or personal computers began to rocket in the early 1980s. In the United States, for example, sales of personal computers soared from only about 35,000 in 1980 to nearly 3 million in 1982.

Computers are one of the miracles of the modern ELECTRONICS industry. They are capable of such fantastic feats that they have often been called electronic brains. But they are not intelligent in the human sense. Their "brains" are not capable of independent thought like the human brain. They can only do what they are instructed to do.

Above: A simple computer programmed to help children with mathematics. They can also play a game on it.

Bottom right: Computer graphics have become an invaluable tool in science and engineering. This three-dimensional colour image was produced by a computer at CERN, the European Centre for Nuclear Research in Geneva. It shows the "events" occurring in a detector during collisions between atomic particles.

Below: A powerful microcomputer of elegant design, with an integral video display unit. It incorporates twin disk drives, and its random access memory (RAM) can be expanded up to 896K.

Number Crunching

Calculating the payroll of a large company requires that millions of arithmetical calculations have to be carried out. For each person the number of hours worked, rate of pay, deductions and allowances are different. In the past calculating the payroll was a laborious job performed by armies of clerks. Today it can be done in a trice by a central computer, because manipulating numbers – "number crunching" – is what a computer does best. The biggest computers can perform hundreds of millions of arithmetical operations every second!

Data Processing

A weather computer illustrates another strength of the computer – as a data processor. Meteorologists feed into it weather information (data) in the form of figures giving rainfall, temperature, humidity, wind speed and direction from sites all over the world. The computer stores all the information and can sort it out to show any patterns in the way the weather is developing. It can then make predictions about the future weather.

Another main function of the computer is as a controller – as in the driverless train. It is now used to control a host of industrial processes, from machining metal to oil refining, and provides the key to automation (see page 65). The computer is a good controller because it can absorb and analyse information, and act upon it, very quickly.

Computers are playing an ever-increasing role in the process of improving the flow of information between people and machines. This is a field now becoming known as information technology (IT).

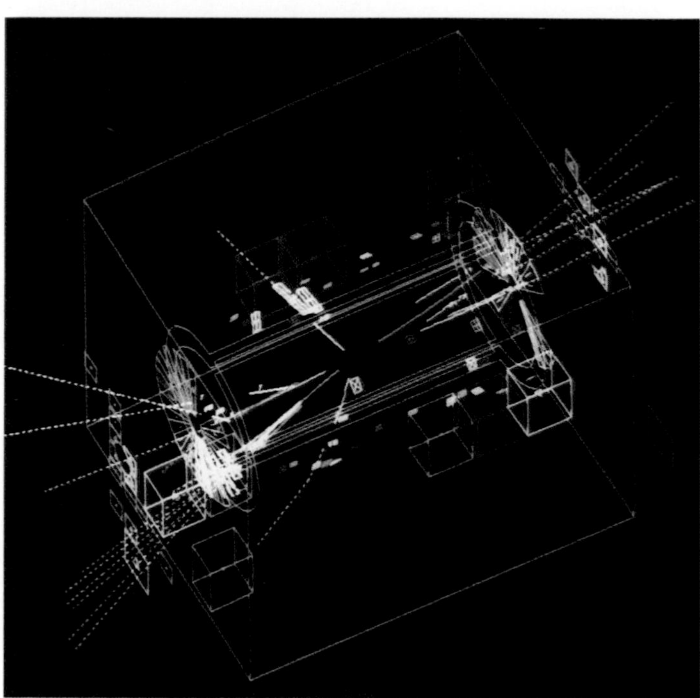

Computers 2

Hardware and Software

A computer system is generally considered to be made up of two parts – the hardware and the software. The hardware consists of the physical parts of the machine. The software consists of the information and the instructions given to the computer that enable it to operate. The information is called the data, and the set of instructions is called the program.

The most dramatic change in the computer world in recent years has been in the shrinking size of the hardware. A modern desk-top computer the size of a typewriter has more computing power than a room-size computer 20 years ago. This has been made possible by the introduction of the silicon chip (see page 144).

Computers vary widely in their capacity for storing and processing data. The most powerful ones, termed mainframes, are used by big business and government organizations. The smallest are the microcomputers used in the home, which have at their heart a single silicon chip called a microprocessor.

Binary Code

Before data and instructions can be fed into a computer, however, they must be coded into numbers, or digits. Because the computer handles information in the form of digits, it is called a digital computer. There is another type of computer, called an ANALOG COMPUTER, which works in a different way.

The digital computer works not with the ordinary decimal digits, 0–9, but with just the two digits, 0 and 1. These two digits can be represented in its electronic circuits by the flow (1) or non-flow (0) of electricity. This can be achieved by switching electric current on or off. And, simply stated, computers do consist of thousands upon thousands of electronic switches.

A number can be expressed in terms of 0s and 1s in the so-called BINARY NUMBER system. In this system place values go up in powers of 2, rather than in powers of 10 as they do in the decimal number system. In computer terminology, each 1 and 0 is called a bit (short for "binary digit"). And each number, letter and symbol is expressed as a code of eight bits. The eight-bit unit is called a byte.

Programming

Computers perform phenomenal feats of calculation, but they do not do so in a complicated way. They actually carry out very simple operations, such as addition and subtraction. They achieve their fantastic computing power by carrying out these operations at incredible speed. The program, or set of instructions for operating the computer, is therefore written as a sequence of very simple steps. They often take the form of a question that requires a Yes

```
5 REM*** MOTORIST'S MONITOR ***
10 REM*** BOB CHAPPELL *** 26/1/83 ***
15 CR$=CHR$(13):LD$="0/0/0":DIMP$(50),P(50)
20 FORJ=1TO50:P$(J)="*":NEXT
25 REM*** IS THIS FIRST RUN ***
30 PRINT"C       MOTORIST'S MONITOR"
35 PRINT:PRINT"PLEASE ENTER THE DATE IN ANY FORMAT"
40 PRINT:INPUT"YOU WISH";CD$
45 PRINT:PRINT"HAVE YOU PREVIOUSLY SAVED ANY DATA"
50 PRINT:INPUT"ON DISK (Y/N)";A$:IFA$="Y"THENGOSUB535:
   GOTO70
55 PRINT:PRINT"WHAT WAS THE ORIGINAL MILEAGE ":
   PRINT:INPUT"READING";BR
60 PR=BR:CR=BR:LD$=CD$
65 REM*** MENU ***
70 PRINT"C       MOTORIST'S MONITOR"
75 PRINT:PRINT:PRINT:PRINT:PRINT" 1.  FUEL UPDATE."
80 PRINT:PRINT" 2.  REPLACEMENTS UPDATE."
85 PRINT:PRINT" 3.  SERVICING/REPAIRS UPDATE."
90 PRINT:PRINT" 4.  STATISTICS."
95 PRINT:PRINT" 5.  CHECKUPS DUE."
100 PRINT:PRINT" 6.  EXIT FROM THE PROGRAM."
105 PRINT:PRINT:INPUT"YOUR CHOICE IS NUMBER";N
110 IFN<1ORN>6THENPRINT:PRINT"SORRY - NO SUCH OPTION.":
    GOSUB620:GOTO70
115 ONNGOSUB130,185,260,320,400,585
120 GOTO70
125 REM*** FUEL UPDATE ***
130 PRINT"C":PRINT
135 PRINT"MILEAGE READING ON ";LD$:PRINT:PRINT"WAS";PR
```

Above: An extract from a computer program devised to help motorists check their expenses.

Right: The computer has found its way into the milking parlour, allowing farmers to keep an accurate check on their cows' milk yield, for example.

Below: An engineer uses a light pen on the video display unit of this 32-bit computer, to modify a design he has prepared.

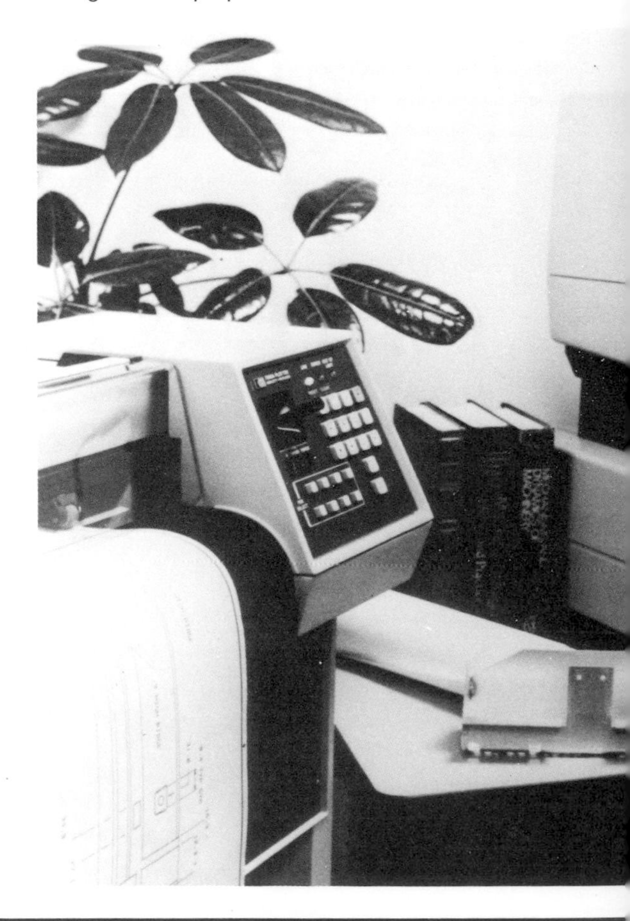

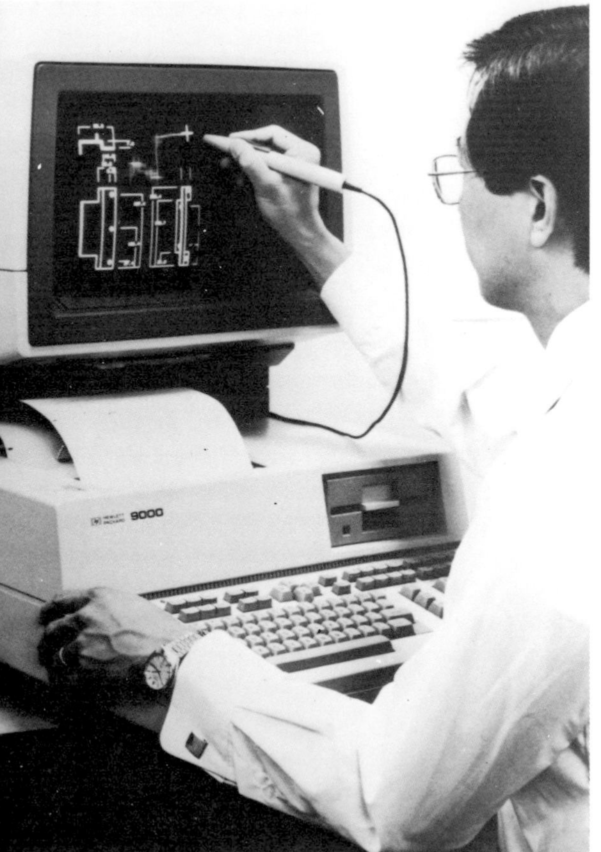

or No answer (such as "Is A larger than B?"), which the computer can represent as an On or Off electric pulse.

Before computer programmers write a program, they prepare a flowchart that breaks down the computing operation into simple stages and shows how these are related.

Fortunately, the programmers do not have to translate the program into binary code. Instead they write it in simplified English in a so-called computer language, which the computer can recognize. Then the computer itself converts this language into binary code. Several computer languages have been developed for different applications, including BASIC, COBOL, FORTRAN and PASCAL. Writing programs is very skilled and time-consuming work. But for most typical computer applications ready-written programs are available.

The Computing Process

The program and data are fed into the computer through an input device. This usually takes the form of a keyboard like that of a typewriter, but with additional keys. Data may also enter the computer from magnetic-tape or magnetic-disk units. The disks may be "floppy" (flexible) or "hard" (rigid). They are played on a disk drive, which is much like a record player. Other input devices include optical readers that can recognize printed patterns (such as the barcode on shop goods), or characters.

The program and data pass into the computer's data-storage circuits, or memory unit. A central processing unit (CPU) then acts upon data according to the instructions in the program. Afterwards it directs the results back into the memory and thence out of the computer through an output device. The CPU is the heart of the computer, which carries out arithmetic calculations and also controls the other units. A built-in clock ensures that all the operations are properly synchronized.

The output device may be a high-speed printer capable of printing several hundred characters a second. Or it may be a video display unit (VDU) like a television screen. Results may be displayed on the VDU numerically or in pictorial form, as graphics. Graphics are made up of combinations of so-called picture elements, or pixels.

The keyboard, disk drives, VDU, and printer are often separate units external to the CPU and are known as peripherals. There are several other kinds of peripherals that expand the capacity and versatility of the computer, particularly of the basic home microcomputer. These include a joystick for playing video games and a modem. The modem is a unit that converts computer signals into audio (sound) signals that can be sent and received over the telephone. There are also music synthesizers and even speech synthesizers, used with appropriate software.

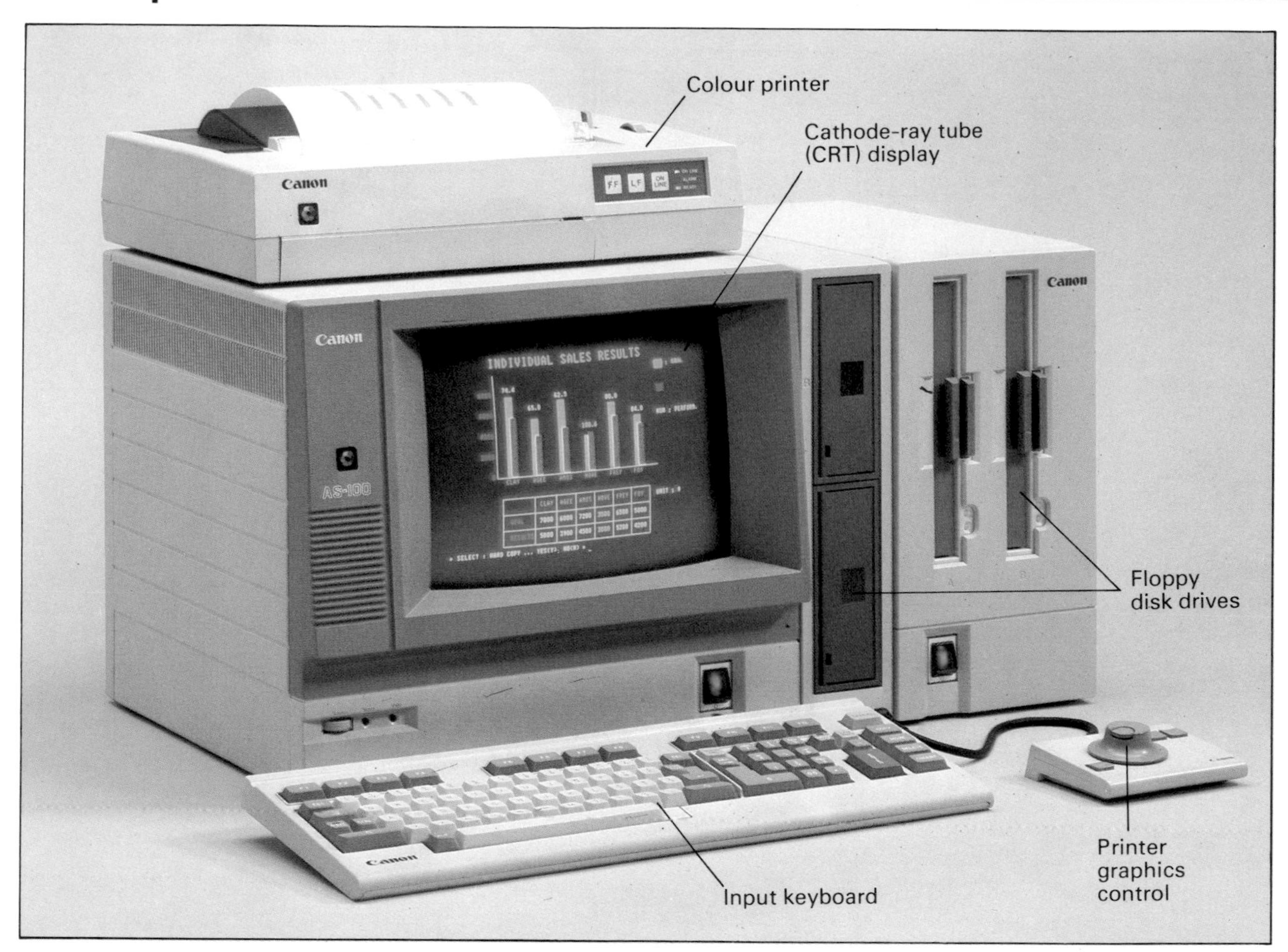

The Memory

The internal memory of a computer consists, like the CPU, of thousands of electronic circuits etched on silicon chips. Particular items of information are stored at specific sites on the chips, called addresses, from where, under instruction, they can be retrieved.

There are in fact two types of computer memory, known as ROM and RAM. ROM stands for "Read-Only Memory". It stores data and instructions essential for starting and operating the computer. It is a permanent memory, preprogrammed by the computer manufacturer, and cannot be altered or erased.

RAM stands for "Random-Access Memory". It contains programs and data for the particular tasks the computer is performing at any time. They are entered into the RAM through an input device at the beginning of the computing process. The RAM can be altered and replaced as necessary, and is lost when the computer is switched off. It is termed "random access" because information can be retrieved at random from any part of the memory.

The power of a computer is usually expressed in terms of the data-storage capacity of the RAM, and of the number of bits of information the computer can operate on at one time. The RAM is usually measured in units of

Above: A Winchester disk-drive unit, which stores data on rigid, rather than floppy disks. The units have a high storage capacity, typically between 5 and 10M.

Left: A 16-bit microcomputer with additional hardware. The colour printer uses ink jets to print in seven colours. The larger of the floppy disk units has a total storage capacity of 2M (megabytes, or millions of bytes).

Below: Computers are invaluable in a busy travel agency to keep track of holidays, prices, bookings, and so on.

Read/write heads

Double-sided magnetic disks

Printed circuits

Disk drive motor

1024 bytes, or kilobytes (K). The first-generation home computers now in widespread use have a RAM of up to 64K. And they are termed 8-bit because they operate with 8-bit units. More powerful 16-bit and even 32-bit machines are also coming onto the market that have a RAM of 256K, 512K or even more. Further memory capacity is provided by back-up units on disk or tape.

Databanks

Small computers may also be able to tap the more extensive memories of larger computers through a cable or modem. This kind of set-up is now common in large companies, where people at small computer terminals ("work stations") can gain immediate access to all the information stored in a mainframe master computer. This has brought closer the day of the "electronic office", where information is handled, mailed and filed electronically, rather than in the form of typed letters on paper.

In some countries a variety of other databanks, or data bases, are available for use by anyone with access to a computer. The United States alone has over 1500 such networks. They provide updated information on a variety of subjects. Some, like Source, provide general information, such as airline schedules, stock market prices and weather forecasts. Others provide specialist services. The American Medical Association operates one called AMA/NET, which includes, for example, information on the latest drugs available. Another, called Agrivisor, provides a farming information and advisory service.

Even without a home computer people in some countries can gain access to computer databanks through a viewdata network. This uses signals transmitted over the telephone to retrieve and display information on a TV screen.

Below: The electronic instrument panel of the MG Maestro car. It incorporates a computer and a voice synthesizer, which "speaks" to the driver. For example, it reminds the driver to wear a seat belt and warns if the battery is not charging.

Atom Smashers

Above: An aerial view of the 3-km long particle accelerator at the Stanford Linear Accelerator Center in California. Built in 1966, it accelerates electrons and positrons to high energies. Positrons are similar to electrons, but they have a positive rather than a negative electric charge. They are the antiparticle of the electron.

From the beginnings of civilization people have tried to answer such basic questions as, how is the universe put together? What is the nature of matter? Thanks to the progress of science and technology, we are at last able to answer these questions satisfactorily.

Our investigations through telescopes and via space probes have revealed a universe of enormous size, which grows more fascinating with each new discovery (see page 164). It is populated not only with stars but with many kinds of other strange bodies. Our investigations into matter have revealed an equally fascinating "inner universe" populated with an assortment of strange particles.

Splitting the Atom

All matter is made up of tiny particles called atoms. Each of the chemical elements – iron, oxygen, chlorine, sodium, and so on – is made up of a different kind of atom. Chemical compounds, such as sodium chloride (common salt), are formed when different atoms combine. It was once thought that atoms were the smallest particles of matter – the word "atom" means "that which cannot be divided".

However, over the past century scientists have shown that the atom is made up of a host of even smaller particles. It is rather like a kind of miniature solar system, having a central nucleus ("Sun"), surrounded by a number

Top right: Tracks of atomic particles photographed in a bubble chamber. The particles are far too small to be seen, but they can affect the matter they pass through. In a bubble chamber they cause a stream of tiny bubbles to form in their wake, and this can be photographed. This photograph shows the effects of collisions between the incoming particles and atoms in the bubble chamber.

Right: A physicist at CERN in Geneva examines tracks photographed in a bubble chamber. By measuring the kinds of tracks produced, she can determine the nature of the particles that made them.

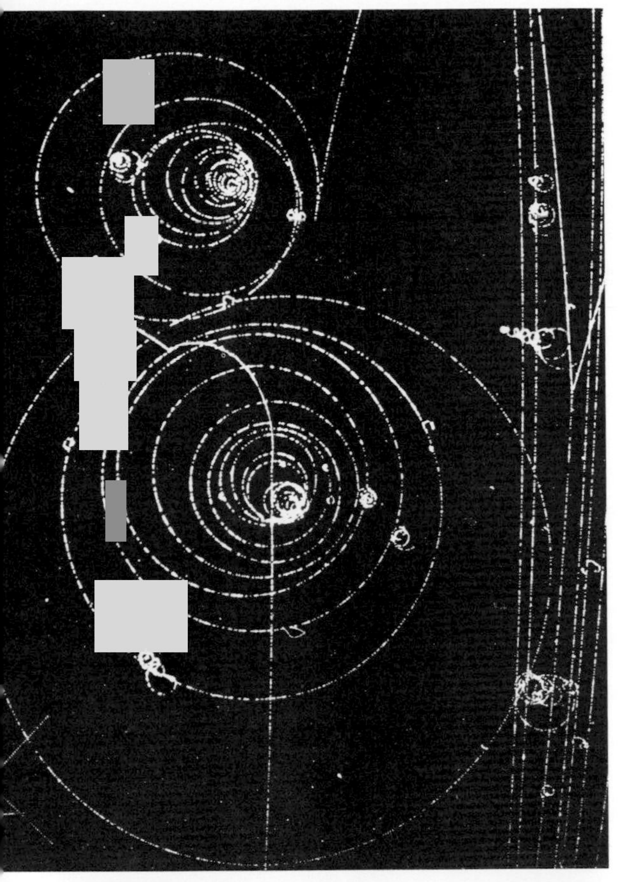

of orbiting particles called electrons ("planets"). The nucleus in turn is made up of two main particles: protons, which have a positive electric charge; and neutrons, which have no charge. There are as many electrons, which have a negative electric charge, as there are protons. And so the atom as a whole is electrically neutral.

Some atoms, such as those of radium, are unstable. They continually break down into other atoms by giving out particles or radiation or both. Some atoms, notably uranium, are so unstable that they sometimes completely split apart. This fission process is made use of in nuclear power stations (see page 18). Other atoms can be made to split by bombarding them with atomic particles. The bombardment also releases other types of particles.

Nearly 200 of these so-called ELEMENTARY PARTICLES have been detected. Some, such as electrons, are considered true elementary particles. But others, such as protons, are now thought to be made up of at least four smaller particles called quarks. These each have different properties and have been called the "up", "down", "strange" and "charmed" quarks!

Nuclear Bombardment

To smash the atom to pieces, you must accelerate the particles doing the bombardment to high speed. This is done in machines called particle accelerators, popularly known as atom-smashers. In these machines charged particles, usually protons and electrons, are accelerated by means of an electric field. Atom-smashers work on the simple principle that electric charges of the same kind (for example, two positive charges) repel each other. The same kind of thing happens between two magnets.

There are two main types of accelerators. In the linear accelerator (linac), particles are accelerated in a straight line. To achieve the high speeds necessary, linacs have to be very long. One at Stanford, in California, is over 3 km long. The other type is the circular accelerator. These machines contain powerful magnets that make the accelerating particles move in a circle. The cyclotron was the first circular accelerator, and has been developed into machines of incredible power like the synchrotron.

Two of the most powerful synchrotrons are at the European Centre for Nuclear Research (CERN) in Geneva, and at Fermilab in Batavia, Illinois. CERN's synchrotron consists of a circular tube over 2 km in diameter, located in an underground tunnel. The tube is surrounded by nearly 1000 magnets, which keep the particles in the tube moving in a circle. In operation it consumes as much electricity as a small town. An even more powerful machine, known as Isabelle, is being built at Brookhaven National Laboratory in New York State. It has magnets made from SUPERCONDUCTORS.

Lasers

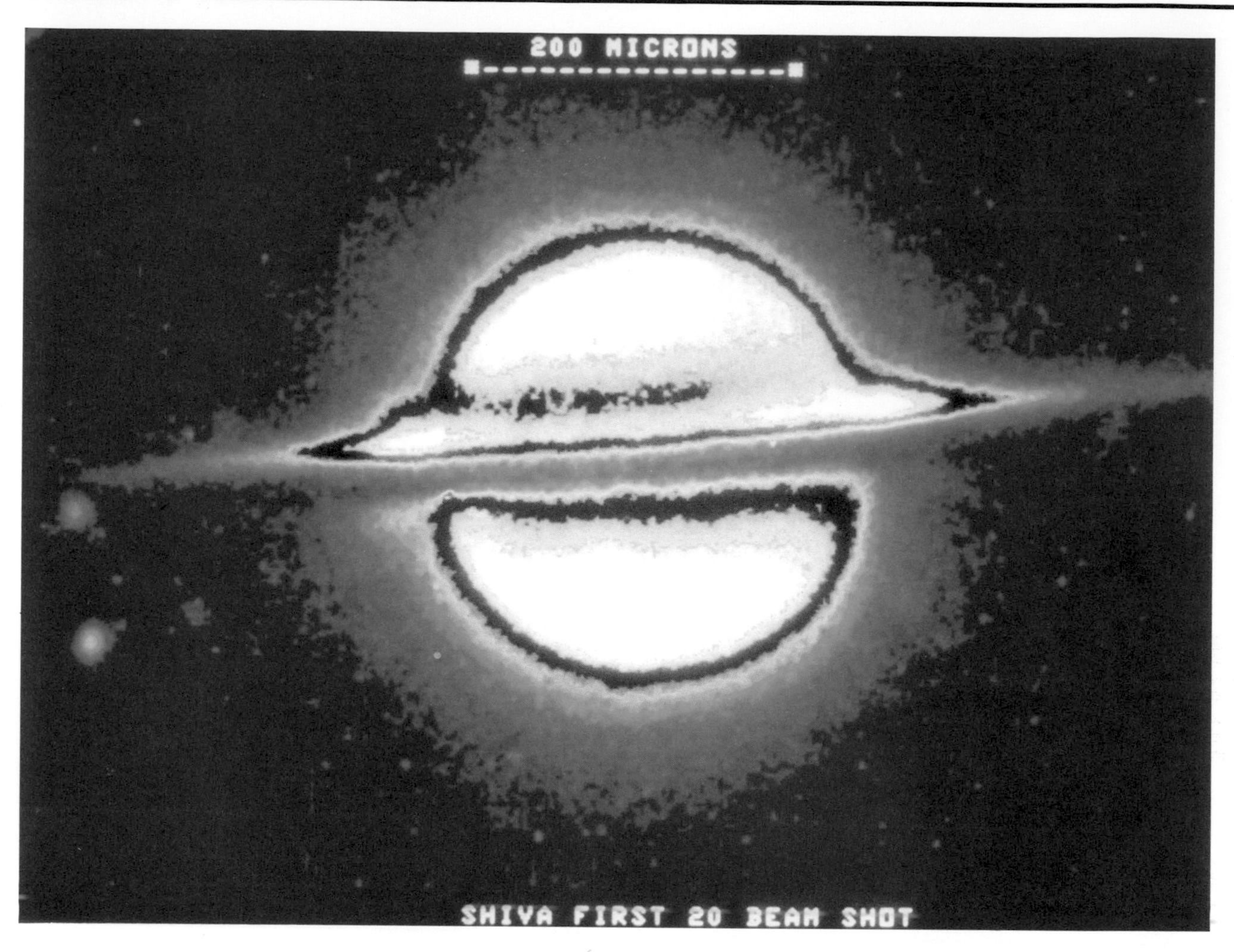

Above: A target pellet being blasted by 20 laser beams during nuclear fusion experiments at Lawrence Livermore Laboratory in California. The laser beams heat the pellet, which is made of heavy hydrogen, to a temperature of millions of degrees. This is high enough for fusion reactions to occur.

No invention of recent years has captured the imagination as much as the laser. Many people think of the laser as a kind of "death ray" that can slice through metal – and flesh – in an instant. And some lasers have been developed as weapons of destruction. But the majority are put to more practical use in industry, communication, medicine and scientific research.

Lasers are electronic devices that produce intense beams of pure light. The word "laser" stands for "light amplification by stimulated emission of radiation". This describes the principle on which it works.

When a laser is operating, its atoms get "excited" and give off radiation. This radiation in turn triggers off, or stimulates, other atoms into emitting more radiation. In this way a build-up of radiation occurs, and eventually a powerful beam emerges. It is visible radiation – that is, light – of a particular wavelength (or colour).

Laser Types

The most powerful lasers are made up of rods of solid material, such as ruby or specially "doped" glass. Solid lasers produce their beam in pulses that can have unbelievable power. One of the most powerful is a glass laser called

Top right: Part of the laser system known as Shiva used in the experiments mentioned above. Each of the tubes in the picture serves as an amplifier to produce one of the 20 powerful beams. Together these beams supply some 25 million megawatts (MW) of power, though for only a fleeting moment. (For comparison, a reasonably large electric power station may produce about 3000 MW.)

Right: The lasers used in laboratory research work are very much lower powered. They are usually gas lasers, which produce a continuous beam. A laser beam can be focused by lenses and reflected by mirrors, just like an ordinary light beam.

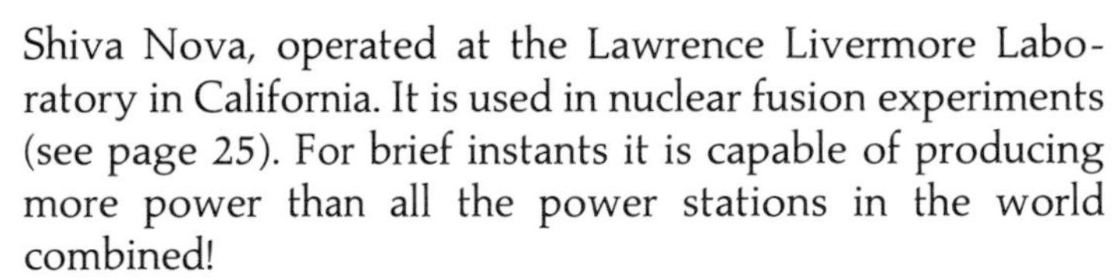

Shiva Nova, operated at the Lawrence Livermore Laboratory in California. It is used in nuclear fusion experiments (see page 25). For brief instants it is capable of producing more power than all the power stations in the world combined!

The other main types of lasers, gas and semiconductor, are not so powerful as the solid type, but can produce a steady power output. The gas laser consists of a column of gas, such as carbon dioxide. The semiconductor laser uses slices of semiconductor material, such as gallium arsenide.

Pure and Parallel

Unlike ordinary white light, laser light is made up of a single wavelength (or colour) and has all its waves in step. Because of this it can be used to carry signals. This is happening in the exciting new development of FIBRE-OPTICS, which is helping to revolutionize communications (see page 139). The purity of the beam also allows the production of true three-dimensional (3D) pictures, in a technique known as HOLOGRAPHY.

Whereas the beam from an ordinary light source fans out broadly, a laser beam stays parallel. This makes it useful for guiding machines that must stick to an absolutely straight course. This precision is often required in tunnelling operations (see page 83). The parallel nature of the beam also makes it easy to focus and control with pinpoint accuracy.

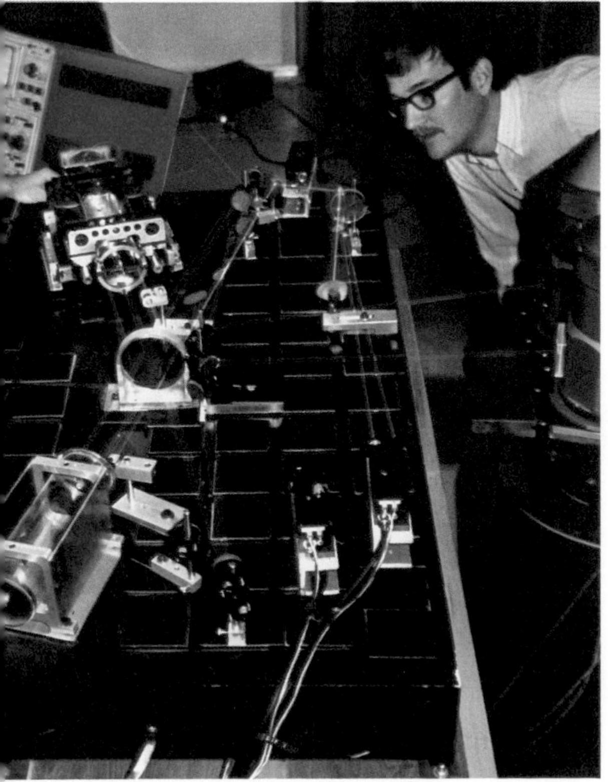

Microwelding

This accuracy is utilized, for example, in delicate welding and precision cutting operations in industry. Eye surgeons have also used the laser for welding. They heat-weld retinas back in place when they are becoming detached inside the eye. If the detachment process were allowed to continue, blindness would result. Eye surgeons also use lasers to punch tiny holes in the eyeball to relieve the pressure that occurs in an eye disease called glaucoma.

The laser has other medical applications too. It can be used to burn away skin cancers, and as a "scalpel" to cut patients open in surgery. The cutting is very precise, and the heat of the laser seals the cut blood vessels, reducing the bleeding.

All this is a far cry from the "death-ray" image of the laser. But military applications of the laser have not been neglected. For example, laser bombing methods have been developed. The target is illuminated by a laser beam, either from the ground or from the air; then the bombs are dropped. They are fitted with sensors that detect and home onto the laser-lit target. Up in space Russia has experimented with high-energy lasers that could be turned into weapons and directed against other spacecraft.

Scientific Instruments 1

We may be the most dominant species on Earth, but we live by our wits rather than by our senses. Compared with certain other animals our senses of sight, touch, taste, hearing and smell are poor. Hawks, for example, can detect a mouse in the grass while hovering 30 metres above the ground. Pit vipers are sensitive to invisible infrared heat radiation, and so can find their prey in the dark. Dogs can hear high-pitched sounds beyond the range of human hearing; hence "silent" dog whistles.

So over the years we have invented all kinds of instruments and devices to extend the range of our senses. And the progress of science and technology has depended a great deal on the development of ever more versatile and accurate instruments.

One of the first devices invented was a pair of scales to measure weight. This was in Egypt before 3500 BC. Later, in Babylon, standard weights came into use. This was an important milestone, because measurements of any kind are useless unless they are related to a fixed standard. In the modern world, therefore, every measurement, from length to brightness, is expressed in terms of very precise units. Scientists use the international system (SI) of measurement units, based on the metre (for length), kilogram (for weight) and second (for time).

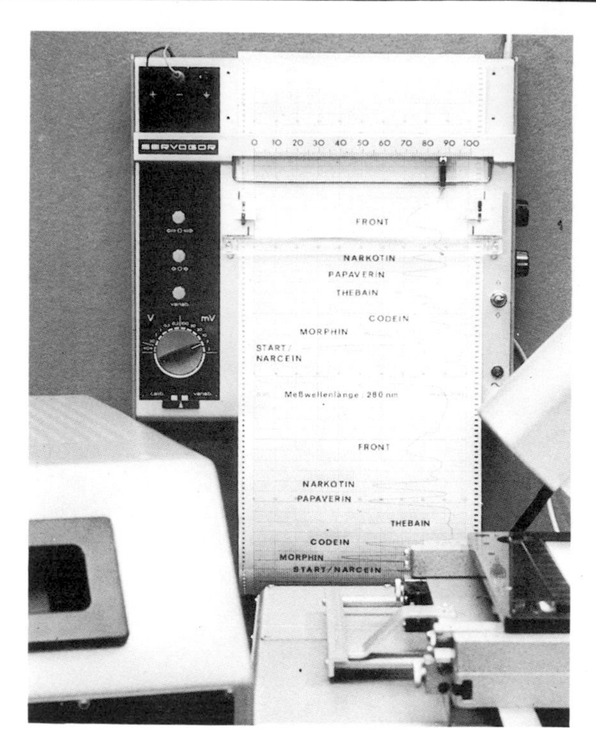

Above: A pen recorder displays the results of chemical analysis as a moving trace on paper. The peaks on the trace indicate the presence of specific chemical compounds.

Below: A scanning electron microscope.

Measuring and Weighing

There are hundreds of different types of instruments in widespread use in science and technology that can conveniently be grouped into those that weigh and measure; and examine and analyse. Measuring instruments form the biggest group. Many of them have names that end in "-meter", which means measurer. The include the thermometer, for measuring temperature; the barometer, for measuring air pressure; the hygrometer, for measuring humidity; the tachometer for measuring engine speed; the ammeter, for measuring electric current.

Other instruments have names that end in "-graph", which implies that they record a measurement in graphical, that is written, form. A SEISMOGRAPH records vibrations in the ground caused by earthquakes as a trace on a moving paper sheet. The POLYGRAPH measures many ("poly") body reactions in a lie-detector test.

Among the commonest quantities that have to be measured are weight, dimensions and time. A precision weighing instrument is the BALANCE. Some balances are so sensitive that they can detect the change in weight of a sheet of paper when a few words are written on it in pencil. A widely used precision instrument for measuring dimensions is the MICROMETER GAUGE, or micrometer calipers. Some can measure dimensions to an accuracy of one-millionth of a metre.

For accurate time measurement, the most widely used

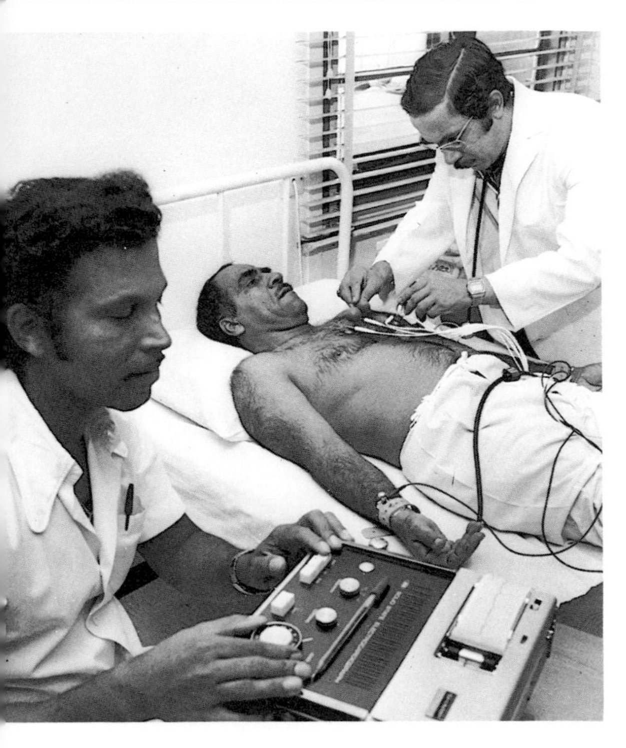

Above: An electrocardiograph in use. This instrument records the faint electrical activity of the heart as a moving trace on paper.

Below: A scanning electron microscope photograph of the claw of a head louse anchored on a human hair (X500).

instrument is the quartz clock, which uses the vibrations of a quartz crystal as regulator. The digital watch (see page 144), a compact version of the quartz clock, has been made possible because the necessary electronic circuitry can now be shrunk onto a silicon chip. Some of the most accurate quartz clocks are kept at astronomical observatories. They are accurate to within about one-fiftieth of a second a year. But even they have to be corrected from time to time by the most accurate clock of all, the atomic clock. This is a complex piece of equipment, looking nothing like an ordinary clock, and is regulated by the vibrations of atoms of ammonia or caesium vapour.

Examining and Analysing

Foremost among instruments that examine and analyse are the microscope and spectroscope which are to be found in all branches of scientific work (see below). The names of many other instruments, of this type end in "-scope", which implies "seeing". The telescope "sees at a distance", and the most powerful models can "see" more than 150,000 million million million kilometres into the depths of space (see page 164). The ENDOSCOPE sees inside things, such as the human body or a turbine. The STROBOSCOPE uses a flashing light to "freeze" the movement of a piece of rotating machinery. The OSCILLOSCOPE uses a cathode-ray tube to display electrical quantities.

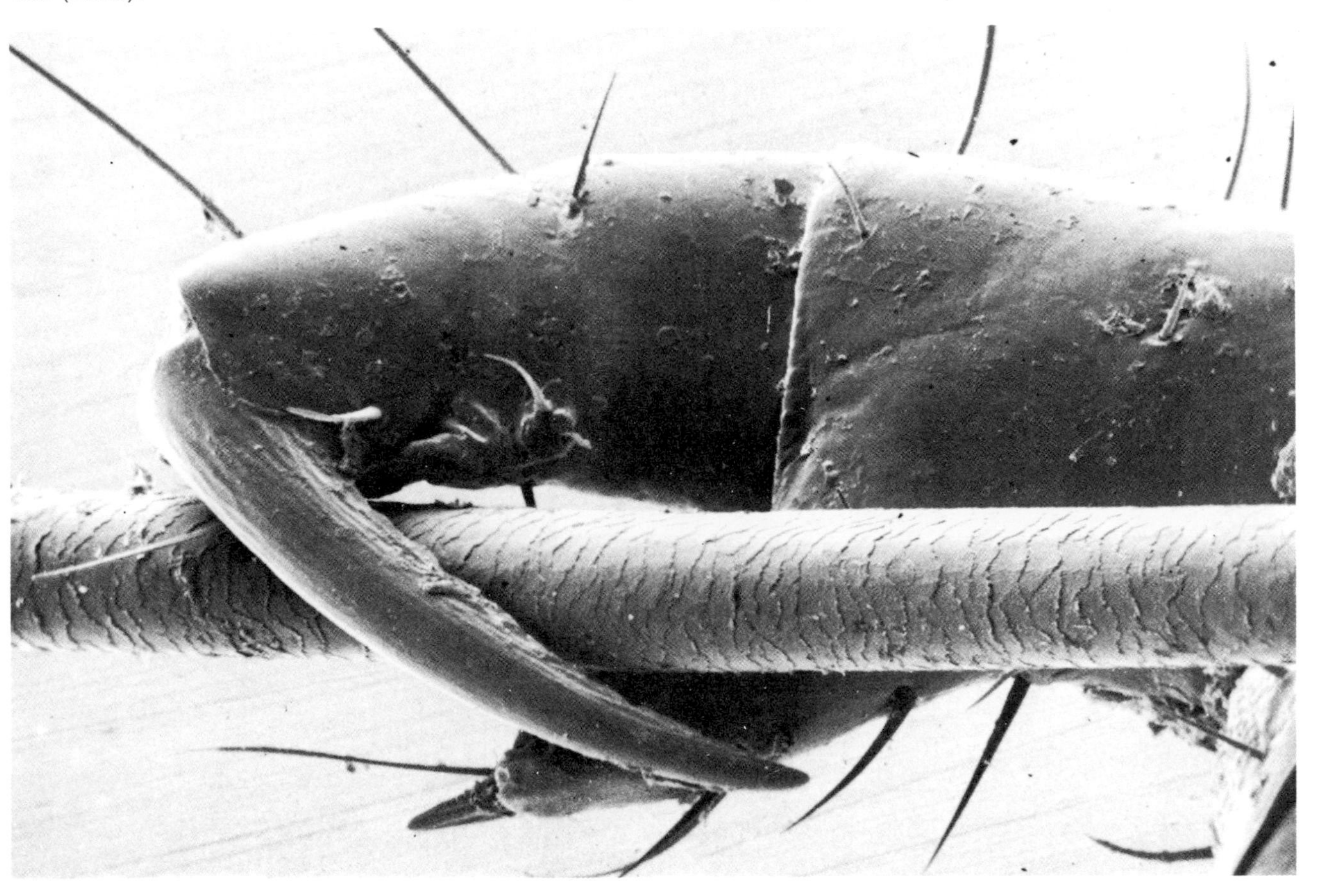

Scientific Instruments 2

Above: In every branch of science electronic apparatus is used for control, analysis and data processing. This equipment is in the control room of the Arecibo radio telescope in Puerto Rico. Note the small cathode-ray tube, or oscilloscope at the top left.

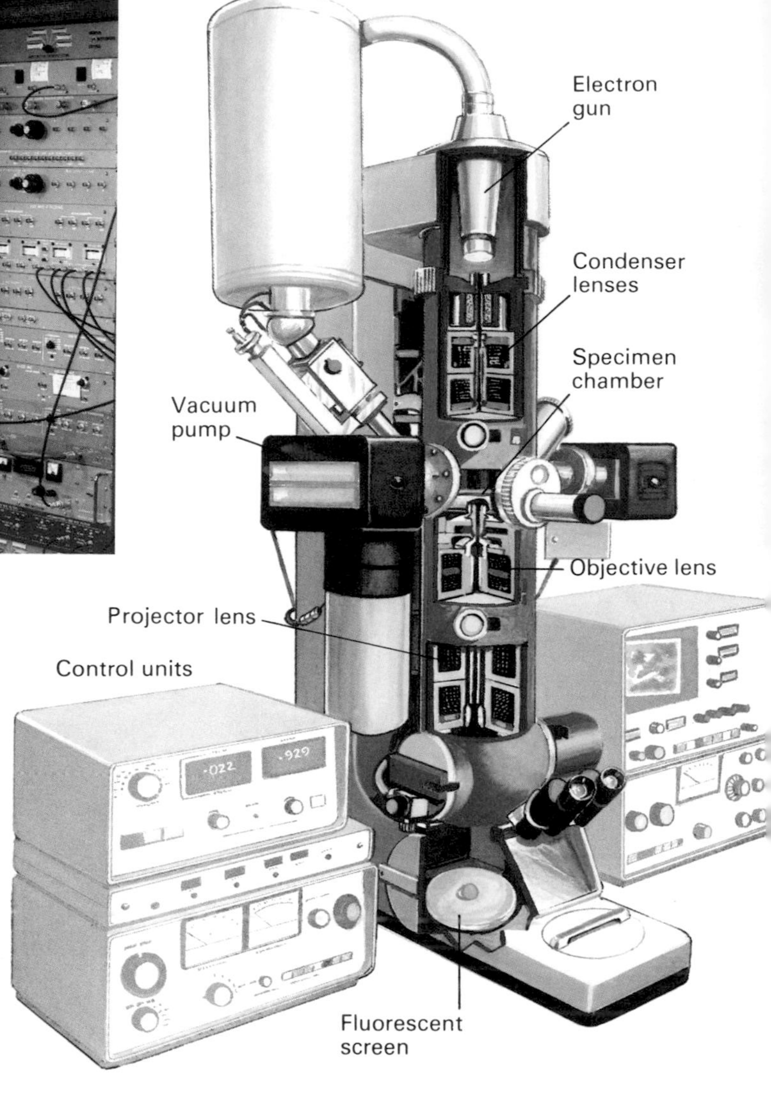

Right: A cutaway of a transmission electron microscope, which fires electrons through thin sections of a specimen. The "lenses" in the microscope are actually magnetic coils, which can focus electron beams just as glass lenses can focus light rays.

The Optical Microscope

The optical microscope is one of the most widely used instruments found in all fields of scientific work, from biology to metallurgy. It uses two sets of lenses to magnify and observe an object, or specimen. The lower lens (objective) of the microscope forms a magnified image of the specimen, which is then observed through an upper lens (eyepiece). The modern instrument has several objective lenses on a rotating turret, which can be brought into operation to change the magnification. It may have a binocular eyepiece and be fitted with a camera.

However, an optical microscope can magnify at most only about 2000 times. For higher magnifications, we must turn to a microscope that "sees" not with light waves but with beams of electrons. Its "lenses" are magnetic coils. An electron microscope can magnify up to a colossal 1,000,000 times. If the full stop at the end of this sentence were magnified one million times, it would increase to over 300 metres across.

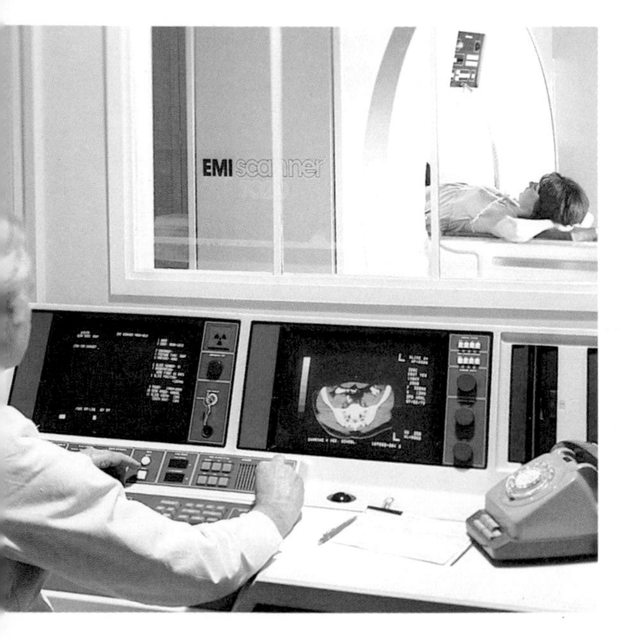

Above: An X-ray scanner at work. In this machine a section of the patient's body is examined from different angles by a thin beam of X-rays. The results go to a computer, which produces a detailed internal view of that section. Body organs can be clearly seen. The proper name for this technique is X-ray tomography.

Below: Very elaborate apparatus is required in the field of particle physics. This picture shows a detection device for spotting chance interactions between atoms and particles called neutrinos. The tubes surrounding the detector are photomultiplier tubes, which are able to detect the faint flashes of light produced during any interactions.

The Electron Microscope

An electron microscope is a very large – and expensive – piece of apparatus, though its operation can be easily explained. At the top is an electron gun which produces a beam of electrons. This is accelerated by means of a very high voltage, focused by magnetic coils, and passed through a very thin slice of specimen. The emerging beam then passes through another coil, which acts like the objective in a light microscope and produces the magnification. An image is formed as the electron beam falls on a fluorescent screen. This is then viewed through a binocular eyepiece, or photographed.

The machine just described is a transmission electron microscope. Another type, the scanning electron microscope, looks at the outside of a specimen by scanning it with the electron beam. It produces rather beautiful, almost three-dimensional images.

The Spectroscope

When sunlight passes through a glass prism, a band of colour is produced which shows "all the colours of the rainbow". It is produced because the prism splits up white sunlight into light of different colours, or wavelengths. This is the principle behind one of the most useful instruments, the spectroscope, or spectrometer.

By studying the spectrum of light given off by a substance, scientists can tell all kinds of things about that substance. From the spectrum of light from a star, astronomers can tell how hot the star is, what it is made of, and how fast it is travelling. Other scientists use the spectroscope to identify substances in chemical analysis and to determine the structure of compounds. These days they not only use light rays to form spectra, but X-rays, gamma-rays and microwaves as well.

One of the most sensitive spectrometers is the mass spectrometer, which forms a spectrum of different substances in a mixture according to their mass. It can detect substances which are present only as one part in a hundred million! This is equivalent to finding a needle in a haystack.

Chromatography

Chromatography is another very sensitive method of analysis now widely used in chemical laboratories to separate and identify substances in mixtures. The word actually means "colour writing", because it was a method originally used to separate coloured pigments. The simplest method uses an absorbent paper. A spot of mixture is dropped onto the paper, which is then dipped into a suitable solvent (dissolving liquid). As the solvent moves through the mixture, it washes out the various substances at different rates, and they eventually separate. The most advanced technique is GAS CHROMATOGRAPHY.

Medical Engineering

Today among the developed nations of the world, people can expect to live at least until they are over 70 years old. Two thousand years ago, they would have been fortunate to reach the age of 40.

This leap in life expectancy has been brought about by many factors: better diet and the improvement in public health; the provision of clean water and the hygienic disposal of sewerage. More particularly it has resulted from the great advances that have taken place in medical science. In recent years they have led to such miracles as test-tube babies and heart transplants, and the complete eradication from the world of the killer disease smallpox.

Advances have occurred in every branch of medicine. Better methods have been found both to diagnose illnesses and diseases (that is, find out what they are) and to treat them. For diagnosis, doctors no longer have to rely solely on experience. They can now call upon the assistance of a variety of equipment to help them. This ranges from the simple stethoscope and sphygmomanometer (which measures blood pressure) to the latest scanning X-ray equipment (see page 159), sonar scanners (right) and ENDOSCOPES. Other sensitive instruments include the ECG and the EEG, which record the faint electrical activity of the heart and the brain respectively.

For treatment they may choose chemical therapy – treatment by chemical drugs (see page 61). They may decide on surgery. This may involve the repairing of damaged tissues and organs, or their replacement. Surgeons may use transplants – organs donated by other patients. Or they may use man-made devices, a technique often called spare-part surgery.

Scientists are now hopeful that one day it may be possible to remove "bad" genes from body cells and stop hereditary diseases. They have already achieved success in removing and transferring genes in lower animals in an exciting new field of study called GENETIC ENGINEERING.

Spare Parts

For centuries people who have lost limbs have been able to replace them with artificial limbs of one kind or another, for example, a wooden leg or a hook for a hand. Only since World War I has the replacement of limbs become more of a science, the science of prosthetics. And an enormous number of people do need artificial limbs of some sort, three-quarters of a million in the West alone. Most of them lost their original limbs in accidents or as a result of disease. But some were born without limbs. In some cases this was caused by the mother taking certain drugs during pregnancy.

Artificial limbs are now tailor-made for the individual and are carefully designed to fit comfortably over the stump, the part of the limb still remaining. Artificial legs

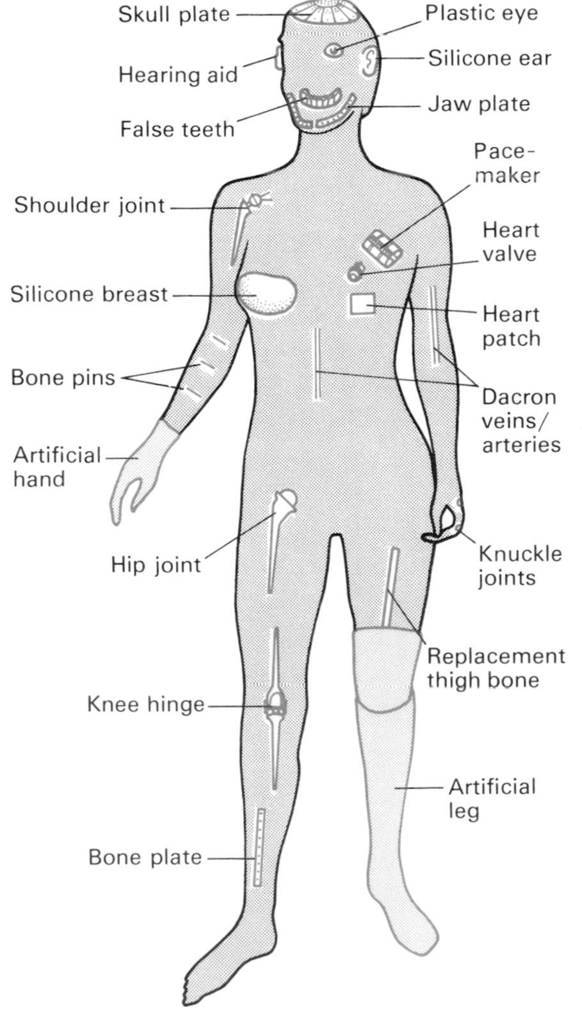

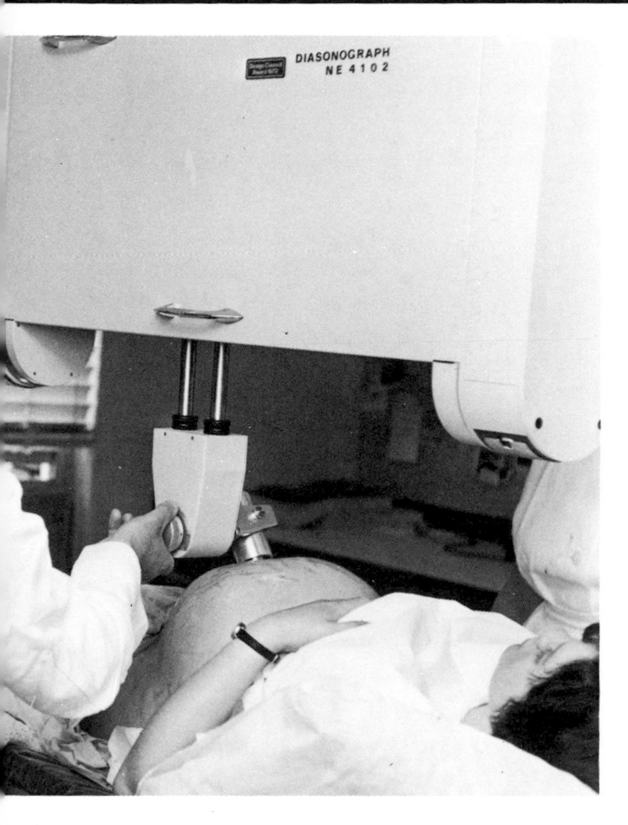

Above: A pregnant woman having a sonar scan. This is a method in which the doctor uses sound waves to "look" inside the womb at the unborn baby. The sound waves are reflected by the body tissues, and the echoes are analysed by computer and displayed as a picture on a screen.

Below: This sonar scan reveals that the woman will have twins.

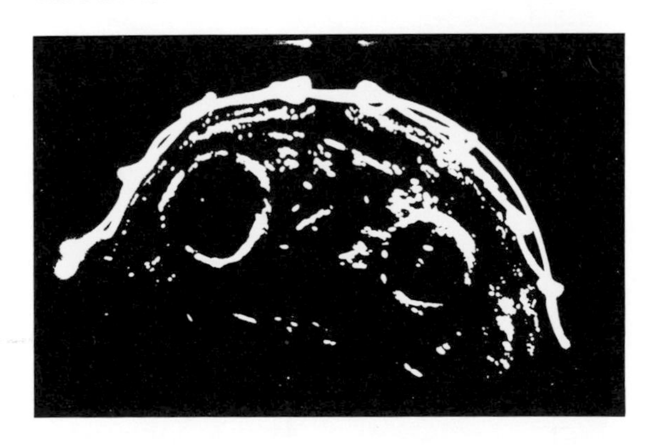

Left: Some of the artificial parts, or prostheses, used to mend the human body when it is damaged by accident or disease.

are generally moved by the person swinging them in a certain way. Some artificial arms are powered by compressed gas. Others work from electrical impulses from the person's own muscles; they are called myoelectric. It is this type that is popularly called bionic. The word is short for "biological electronics".

As well as artificial limbs, people may have artificial eyes made of plastic; hearing aids that fit inside the ear; false teeth. All these are external prostheses. Inside the body, man-made objects may be fitted too, from a metal skull plate to new blood vessels and heart valves.

The plates and pins are made of such metals as stainless steel, titanium or silver, which are very inert materials that do not react with body tissues. Artificial blood vessels, used to rejoin veins and arteries, are made from inert plastics such as Dacron.

Taking Heart

The most common type of heart valve, in use since the 1960s, consists of a hollow chromium alloy ball housed in a kind of cage. The base of the cage is covered in polypropylene fabric, which is sown into the heart tissue.

The natural pumping action of the heart is triggered off by minute electrical impulses from a so-called pacemaker. Sometimes this fails, and the heart cannot pump properly. Again, the medical engineer, or bioengineer as they are now often called, comes to the rescue with an artificial pacemaker. This delivers a regular electric pulse to the heart to keep it pumping. Early pacemakers were worn externally, but the modern ones are designed to be implanted in the chest cavity. They have batteries that do not need replacing for several years. These may be nuclear batteries, which are powered by a small piece of radioactive material. Or they may be very long lasting chemical batteries like the newly developed lithium-iodine cell.

The ultimate solution to a severe heart condition is to replace the heart with another. A heart transplant using a heart from another person is one method now being used with some success. But there are always problems about the body rejecting the new organ, and there could never be enough suitable donors. So the answer lies in the artificial heart. Russian scientists first fitted an artificial heart to a dog in 1937, and since then a variety of devices have been developed.

Some of the most extensive testing of artificial hearts has been at the University of Utah, in the United States. Here artificial hearts have succeeded in keeping cattle and sheep alive for months. In December 1982 their latest type of artificial heart was implanted in a human being in Salt Lake City and kept him alive for 112 days. Called the Jarvik-7, it was made from polyurethane plastic and aluminium and powered by compressed air.

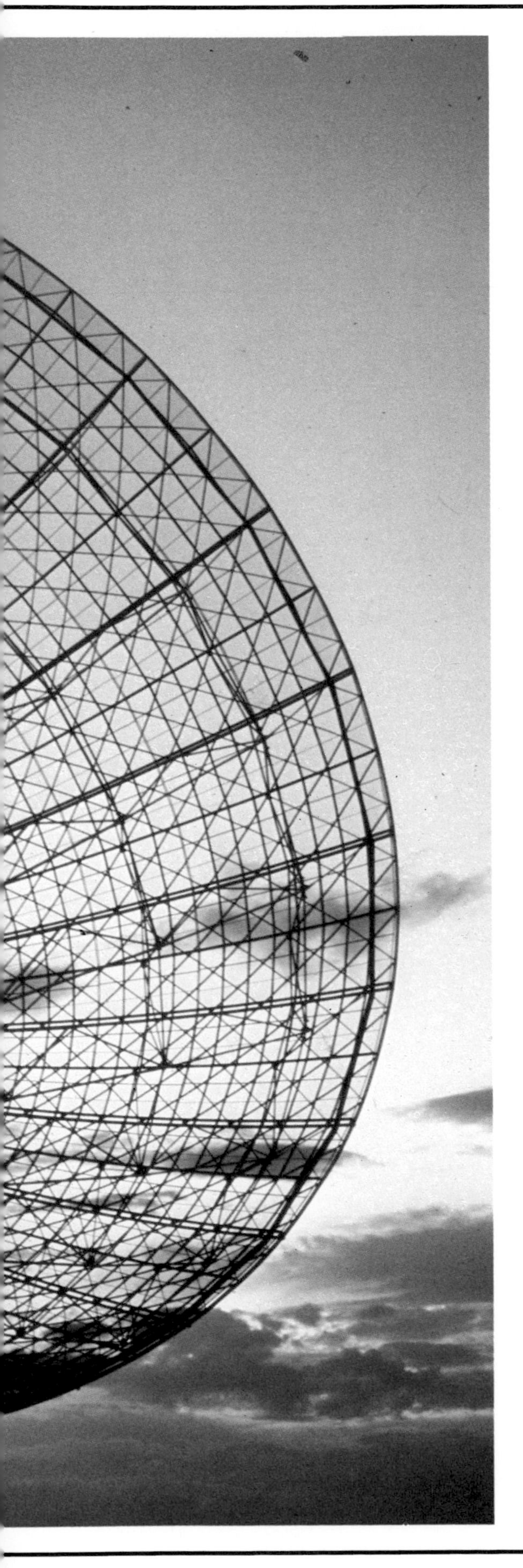

Chapter 9

Space Exploration

Ancient peoples began exploring space, with the naked eye, thousands of years ago. But only after the invention of the telescope in the early 1600s, did astronomers begin to appreciate the true nature of the universe. As telescopes have become more powerful, they have revealed a universe that is much bigger and stranger than anyone can imagine.

In 1957 the modern era of space exploration began with the launching of Sputnik 1, the first artificial satellite. Since then thousands of spacecraft have rocketed into space. Most have become satellites of the Earth; some have ventured to the planets; and some have carried human beings – astronauts.

Astronauts now travel routinely into space, where they can live quite happily for months at a time. They have already journeyed to the Moon; next century they may visit the planets. One day perhaps they may even travel to the stars.

The powerful radio telescope at Parkes Observatory in New South Wales, Australia. Built in 1961, it has a dish 64 metres across. It is used particularly for studying the mysterious distant bodies called quasars.

Telescopes 1

When we look up at the starry heavens, we are looking into the depths of space. As far as we know space goes on and on without end. Space and all the bodies within it – the Sun, the Earth, the planets and their moons – make up what we call the universe or the cosmos. The scientific study of the heavenly bodies is called astronomy.

But without instruments we can see only a few thousand stars. There are many millions more that we cannot see because they are too far away and too faint. To see them we have to look through BINOCULARS or telescopes. These instruments not only gather more light than our eyes, but can magnify also.

The bigger the instruments, the farther we can see into the universe, and the more fascinating that universe becomes. It contains stars without number, which journey through space in great star islands, or galaxies. It contains not only ordinary stars like the Sun, but also stars that explode (SUPERNOVAE); dwarf stars and giant stars. It also contains curious bodies such as pulsars, quasars and black holes.

Observatories

Astronomers carry out their observations at an observatory. For best results, observatories should be located high up in a dry climate. There, the air is cleaner and thinner. For

Above: A panoramic view of Kitt Peak Observatory in Arizona. It is located at an altitude of over 2000 metres. The dome in the foreground houses the 4-metre Mayall reflector.

Below: An aerial view of the Cerro Tololo Inter-American Observatory in Chile.

example, Kitt Peak Observatory in the United States is ideally situated. It is on a mountaintop that rises abruptly from the desert about 60 km from Tucson in Arizona. The skies are perfect for viewing nearly every night of the year.

Kitt Peak and other big observatories, such as Palomar in California, Siding Spring in Australia and Cerro Tololo in Chile, house a variety of telescopes and other instruments to study the stars. Most important among these other instruments is the SPECTROSCOPE, which splits a star's light into a spectrum. Astronomers can tell many different things from the lines in the spectrum. They can tell, for example, whether the star is moving towards or away from us, and what chemical elements it contains.

Professional astronomers do not peer through their telescopes much these days. They take photographs with them instead, exposing the film for long periods. The film "stores" the light coming in and can record very faint objects that the eyes could never see. Photographs are also permanent and can be studied at any time. Photographs of some interesting heavenly bodies appear on pages 168/169.

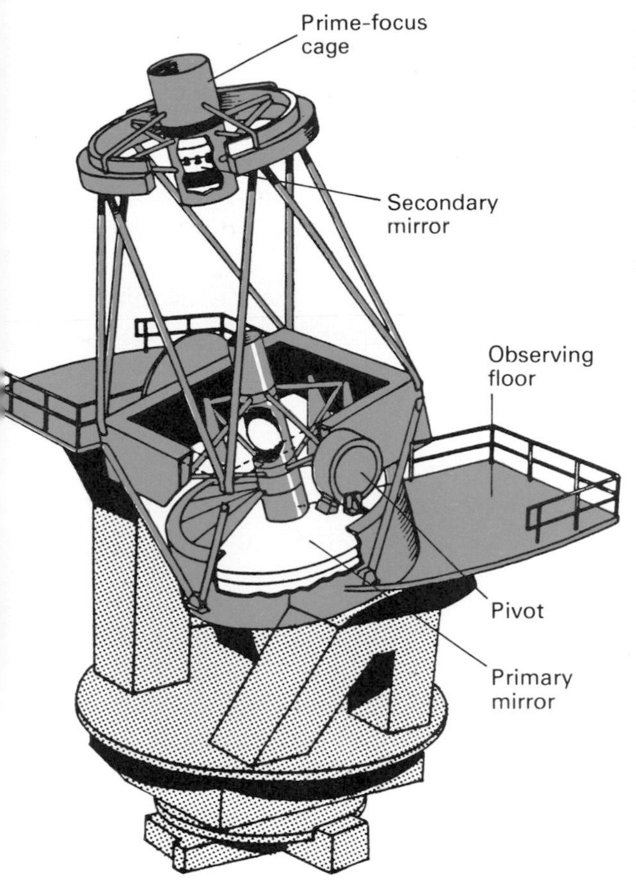

Above: The 4·2-metre Herschel reflector, one of the main instruments at the observatory on La Palma in the Canary Islands.

Below: The 4-metre reflector at the Cerro Tololo Inter-American Observatory in Chile.

Refractors

The first person to look at the heavens through a telescope was Galileo in 1609. He built his telescope with glass lenses – a type known as a refractor. This type is so called because lenses work by refracting, or bending the light passing through them.

A modern refractor has two sets of lenses. The front lens (objective) forms an image, which the rear lens (eyepiece) then magnifies. The lenses are set in tubes which slide in and out of one another so as to focus the image sharply. The image is upside-down, but this does not matter for astronomical work. The largest refractor in the world is at Yerkes Observatory in the United States. It has an objective lens just over one metre across.

Reflectors

Most astronomical telescopes, however, are reflectors. They gather and focus the starlight by means of mirrors, not lenses. The main mirror of a reflector is curved, rather like a saucer. It reflects light onto one or more other mirrors, which in turn reflect the light into an eyepiece. Isaac Newton built the first reflector in 1672.

Reflectors can be built very much bigger than refractors. The biggest has a light-gathering mirror 6 metres in diameter. It has been built (1976) at Zelenchukskaya in the Caucasus Mountains in Russia. It is said to be so sensitive that it can detect the light from a candle 25,000 km away! Other big reflectors include the 5-metre Hale Telescope at Palomar Observatory and the 4-metre Anglo-Australian Telescope at Siding Spring.

Telescopes 2

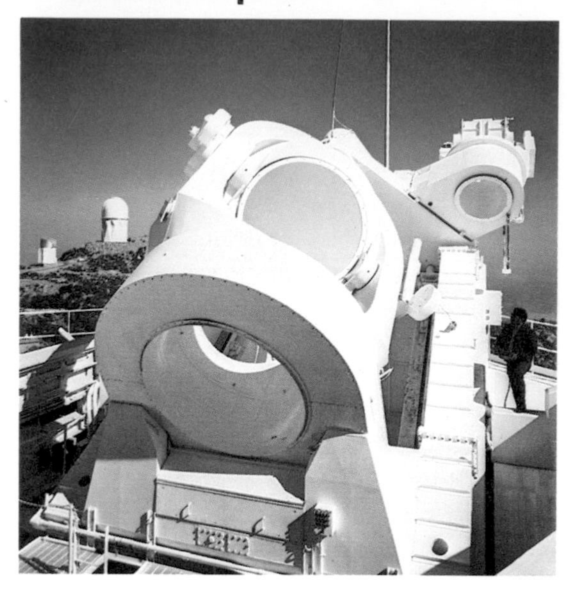

Above: Kitt Peak Observatory in Arizona has some fine telescopes. This photograph shows the mirror of the McMath solar telescope, used for studying the Sun. The mirror, which is 2 metres across, reflects an image of the Sun down to an underground observing room.

Bottom right: The giant radio telescope near Arecibo, in Puerto Rico. Completed in 1963, it is still the world's largest. The dish, which measures 305 metres across, collects radio waves from outer space and focuses them on the antenna high above it.

Below: A view from beneath the dish of the Arecibo telescope. The dish is suspended over a natural bowl, and is made up of 38,778 perforated aluminium panels.

Radio Telescopes

Although reflectors can be massive, they are dwarfed by another type of telescope – the radio telescope. This gathers, not light rays, but radio waves. Stars when they shine give out radio waves as well as light rays. Some of the most exciting discoveries in astronomy in recent years have been made using radio telescopes, including the discovery of QUASARS and PULSARS.

Most radio telescopes are huge metal dishes mounted so that they can be steered and pointed at different parts of the sky. They gather the faint radio signals and feed them to an antenna (aerial) above the dish. The signals are then amplified (strengthened) and fed to computer equipment, which records and analyses them.

The largest steerable radio telescope is at Effelsberg, near Bonn in West Germany. It has a dish 100 metres across. There is an even bigger radio telescope near Arecibo on the Caribbean island of Puerto Rico. It has a fixed dish no less than 305 metres in diameter, built over a natural bowl in a mountaintop.

Some radio observatories use an arrangement of several small dishes to collect radio waves from the heavens. This set-up is as sensitive as a single, very large, dish. The world's largest multi-dish telescope, the Very Large Array in New Mexico, uses 27 movable dishes.

In 1974 the Arecibo telescope was used as a radio transmitter to beam a coded message into the heavens. This message is now travelling towards distant stars, telling that life exists on Earth. Perhaps one day someone "out there" will pick up the message and know that they are not alone in the universe.

Telescopes in Space

As well as light rays and radio rays, stars give off other radiation, such as X-rays, gamma-rays and ultraviolet rays.

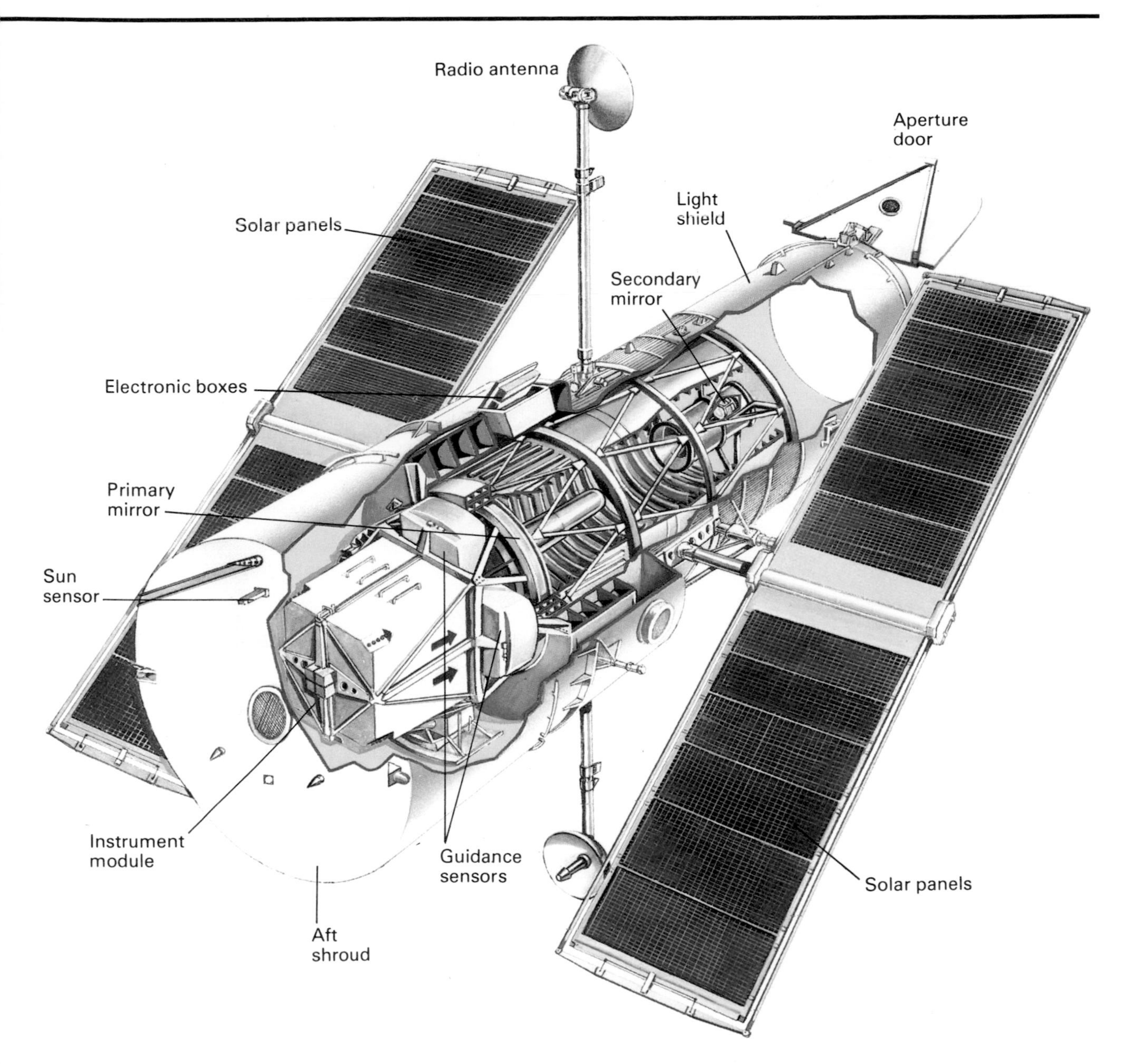

We cannot detect these rays here on the ground. The atmosphere absorbs them. But we can now send instruments up into space to detect them, on satellites. When viewed at other wavelengths, the universe looks very different from the way it does in visible light.

Astronomy satellites such as Copernicus and Einstein have made many amazing discoveries. They have discovered stars called bursters, which give out bursts of X-rays of unbelievable energy. They have also found evidence of BLACK HOLES, awesome bodies that swallow up everything near them, even light.

Astronomers are now waiting to see what startling new discoveries NASA's space telescope (illustrated above) will make. This space observatory has a 2·4-metre diameter reflector, which is able to detect very much fainter objects than we can from Earth.

Above: NASA's space telescope, designed for launch in the space shuttle. It measures over 13 metres long and weighs about 11 tonnes. In its 500-km high orbit it will be able to view the universe very clearly and detect heavenly bodies too faint to be seen from Earth. From time to time astronauts will visit the space telescope to carry out maintenance or bring it back to Earth for repair.

The Heavens

The photographs on these two pages show just a few of the spectacular sights the heavens have to offer. The universe is full of millions upon millions of bodies like these.

The Solar System

Our own little corner of the universe is dominated by the huge ball of glowing gas we call the Sun. There are nine planets that circle the Sun. Our own planet, Earth, circles it at a distance of about 150 million km. The word "planet" means "wanderer", and we can see the planets like bright stars "wandering" across the heavens.

The planets Mercury, Venus, Mars and Pluto are rocky, like the Earth. Jupiter, Saturn, Uranus and Neptune are made up mainly of hydrogen gas. Most of the planets have smaller bodies (satellites, or moons) circling around them. The planets and their moons are the main parts of the Sun's family, or solar system. The Earth has only one satellite – the Moon; Saturn has at least 22.

Earth's Moon lies about 385,000 km away and is our closest neighbour in space. It is a barren, lifeless lump of rock, covered with high mountain ranges, vast plains and huge craters.

The Fixed Stars

The stars lie a long way beyond the solar system. Even the closest star (Proxima Centauri) is over 40 million million km away. This is so far that its light, travelling 300,000 km every second, takes over four years to reach us. We say that this star lies over four light-years away. The light-year is a useful measuring unit in astronomy.

The stars are great globes of glowing gas, like the Sun. They appear small only because they are so distant. Some are in fact hundreds of times bigger and brighter than the Sun. Most stars do not travel through space alone, but with one or more companion stars. Occasionally thousands upon thousands of stars collect together into a massive globular cluster. The space between the stars is not totally empty, but contains grains of dust and gas. Sometimes this interstellar matter forms a visible cloud, or nebula.

Also, the stars are not scattered evenly in space. They are gathered loosely together in great galaxies. The Sun and all the stars we can see in the sky belong to the Milky Way Galaxy. Our Galaxy is typical of many in the universe. It contains about 100,000 million stars, formed into a spiral shape like an enormous pinwheel.

On an even larger scale, the galaxies themselves gather together into clusters. In 1982 astronomers found a supercluster containing millions of galaxies that stretches for at least 700 million light-years. This is a distance of 7000 million million million km! This supercluster is the biggest object yet known in the universe.

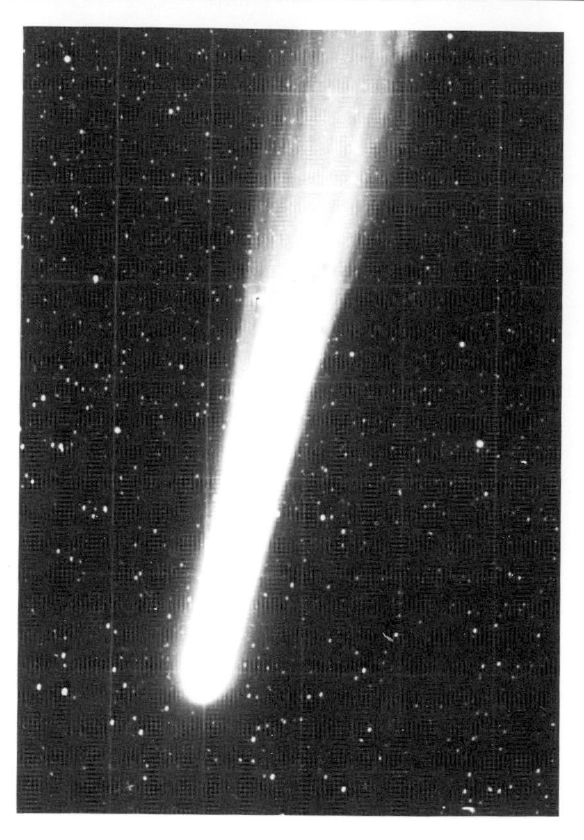

Above: Halley's comet, when last seen in 1910. It will come close to Earth again in 1986.

Below: Man on the Moon. Apollo 17 astronaut Harrison Schmitt examines a lunar boulder in December 1972.

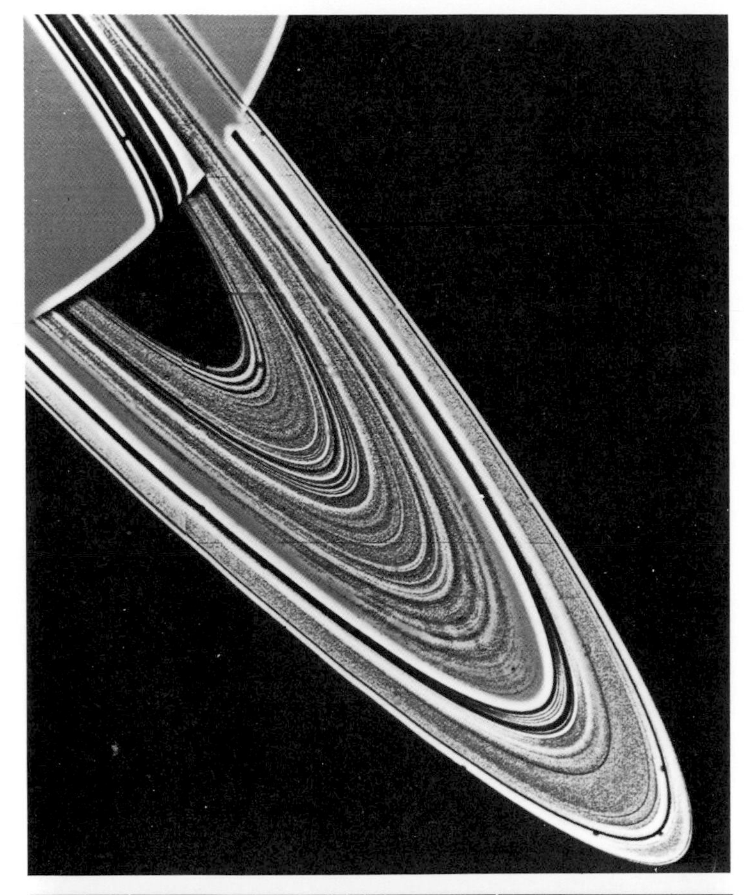

Above: A dense globular cluster in the constellation Hercules. It contains about 150,000 stars.

Left: The spectacular rings around Saturn, which are composed of hundreds of separate ringlets.

Left: A fine spiral galaxy in the constellation Triangulum. Our own galaxy would look like this from a distance.

Below: A cluster of galaxies in the constellation Hercules, some 550 million light-years away.

Rockets and Satellites 1

Above: A false-colour photograph of Italy, taken by the Earth-survey satellite Landsat. Satellite photographs are a boon to map-makers.

Opposite: The launch of the space shuttle, now a regular event but still incredibly spectacular.

Below: The shuttle's European rival, Ariane, rises from the launch pad at Kourou, in French Guiana.

For centuries people on Earth have dreamed of travelling in space. But no one knew whether it would be possible – until 4 October 1957. On that day the Russians launched a powerful rocket that carried a small sphere of aluminium into space. And it stayed there circling round and round the Earth, in orbit.

This sphere, called Sputnik 1, became an artificial satellite, or moon, of the Earth. Tiny though it was, it showed the way ahead. It launched mankind into the Space Age. Today, a hundred or more satellites are launched into space each year, bringing us many benefits.

How do we get an object into space? We have to give it enough speed so that it can overcome the very strong pull of the Earth's gravity, which normally keeps everything down on the ground.

The speed we must give an object to go into orbit is a colossal 28,000 km per hour. This is ten times faster than the fastest supersonic fighter plane! So we need a very powerful engine for propulsion. We also need an engine that will work in space. Jet engines are not powerful enough; neither will they work in space, because they take in air to burn their fuel. The only engine that is powerful enough and will work in space is the rocket.

Rocket Engines

The Chinese discovered how to make rockets nearly 1000 years ago, not long after they invented gunpowder. The rockets they made were much like the firework rockets we enjoy today. These rockets contain gunpowder in a tube, with a hole, or nozzle, at one end. When the gunpowder burns, hot gases are produced which shoot backwards out of the nozzle as a jet. As the gases shoot out backwards, the rocket is thrust forwards, as in a jet engine (see page 108).

Rockets used for space launchings do not burn gunpowder, but a variety of other materials. Such materials are called propellants. Some propellants are solid, others are liquid. Propellants provide not only fuel, but also oxygen to burn the fuel. That is why rockets can work in space, where they is no air.

Solid propellants contain a fuel mixed with synthetic rubber. The most widely used liquid propellants are liquid hydrogen and liquid oxygen. These are used for the main engines of the American space shuttle.

Rockets that burn liquid propellants are more complicated than solid-rocket engines. They have separate tanks to hold their two propellants and a separate chamber for combustion (burning) to take place. They also require pumps to pump fuel from the tanks into the combustion chamber. By burning several tonnes of propellants every second, liquid rockets can produce a thrust of hundreds or even thousands of tonnes.

Rockets and Satellites 2

Step Rockets

But even the biggest single rocket is not powerful enough to go into space by itself. It must be linked up with other rockets, which give one another a "piggy-back" ride into space. A launch vehicle made up in this way is called a step rocket. Usually three rockets, or stages, are linked together. The object to be carried into space, called the payload, is carried in the third, top stage. The European space launching rocket Ariane, developed mainly by France, is a three-stage rocket.

The first stage, called the booster, fires and lifts the vehicle high into the air. When its fuel runs out, it separates and falls away. Then the second stage fires and lifts what remains of the step rocket higher still. In turn this stage falls away. The third stage then fires and carries the payload into orbit.

The payload may be one or more satellites. These are made from lightweight aluminium and magnesium alloys. They carry a variety of instruments, such as cameras and sensors to detect and measure different kinds of radiation, such as X-rays and cosmic rays. They have a tape recorder to record instrument readings, and a radio to transmit the readings back to Earth. And they carry panels of solar cells, which make electricity to power the instruments.

Once satellites are in orbit, travelling at 28,000 km per hour (orbital velocity), they remain there for a long time. Up in space there is no air to slow them down. There are in orbit at present over 4000 "satellites" of one kind or another. Only a few hundred of them, however, are working satellites. Some are satellites that no longer work. And many are simply burnt-out rocket stages – space junk.

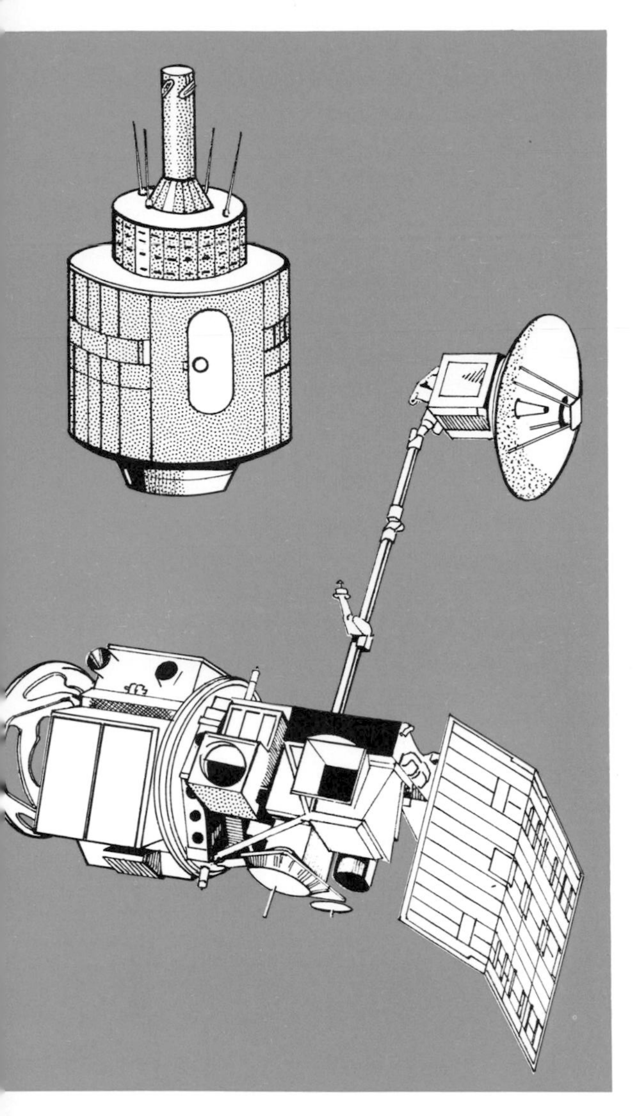

Satellite Servants

Satellites can do many useful jobs in space. We have benefited most in two main areas – weather forecasting and communications. Satellites have also made astonishing discoveries in astronomy (see page 166), and improved map-making and prospecting for minerals.

We cannot as yet control the weather, but we can organize our lives better if we know in advance what the weather will be like. And that is where weather satellites have proved so useful. From their high vantage point, hundreds of kilometres above the Earth, they can look at the weather over a wide area. And they can trace how weather patterns are developing. Armed with this information, meteorologists can issue accurate forecasts.

Some weather satellites orbit a few hundred kilometres high over the poles. They pass rapidly around the Earth about every $1\frac{1}{2}$ hours. Others, such as Meteosat, are located nearly 36,000 km high over the equator. They orbit the Earth every 24 hours. But the Earth itself spins round once in this time. So the satellite appears from Earth to be stationary. We say it is in a stationary orbit. In this position it records the weather over a whole hemisphere.

Most communications satellites, such as the Intelsat Vs, are also positioned in stationary orbit over the equator. From their "fixed" position, they relay signals between countries and continents in the hemisphere they cover. Signals are beamed up to them from one ground station. They receive and amplify (strengthen) the signals and then beam them down to another ground station. The signals handled by the satellites may represent telex messages, television programmes, or telephone calls (see page 136).

Above: Two very successful spacecraft, Landsat 4 (bottom) and Meteosat (top). Landsat surveys the Earth in light of various colours and can pick out details invisible from the ground. Meteosat is Europe's main weather satellite.

Left: This giant Baker-Nunn camera is used to track satellites orbiting high above. It is located at a tracking station in Woomera, near Adelaide in South Australia.

Far left: The launch of a Russian Soyuz spacecraft from the cosmodrome at Baikonur in Central Asia. Russian launch vehicles are recognized by the cluster of four booster rockets around the base. Like the other rockets of the launch vehicle, they burn liquid oxygen and kerosene as propellants.

Right: A photograph returned to Earth by Meteosat 2, in stationary orbit 36,000 km above the equator. It shows the cloud cover over Europe and the Atlantic.

Space Shuttle 1

Below: A cutaway drawing of the orbiter. It is shown launching a satellite with its remote manipulator arm. The orbiter is constructed in much the same way as an ordinary aeroplane, from high-strength aluminium alloys. The structure is protected against overheating by silica tiles and other insulation. The orbiter measures 37·2 metres long and has a wing span of 23·8 metres.

Until November 1982 all satellites were launched into space by a step rocket, which could be used only once. The various bits of the rocket were either lost in the sea or remained in space. In November 1982, however, two satellites were launched in orbit by the American space shuttle. Launching satellites is one of the space shuttle's main jobs.

The space shuttle is different from earlier rocket launchers because it can be used again and again. The main part of the system is the orbiter, shown below. This carries the astronaut crew and the payload, or cargo. The first orbiter, called *Columbia*, made its maiden voyage into space on 12 April 1981. It returned to space four more times over the next 19 months.

The second orbiter, *Challenger*, made its space début in April 1983. Within a few years two more orbiters, *Discovery* and *Atlantis*, will become operational. By the 1990s they will be shuttling into space every two or three weeks. And there will always be an orbiter ready for a space rescue mission in emergencies.

Cheap and Versatile

The space shuttle has several advantages over ordinary rocket launchers. Because it can be re-used, each of its flights is relatively cheap. It can carry up to 30 tonnes of cargo, measuring up to 18 metres long and 4·5 metres in diameter. Another advantage is that it is a manned system. The crew are able to check that the satellites they launch

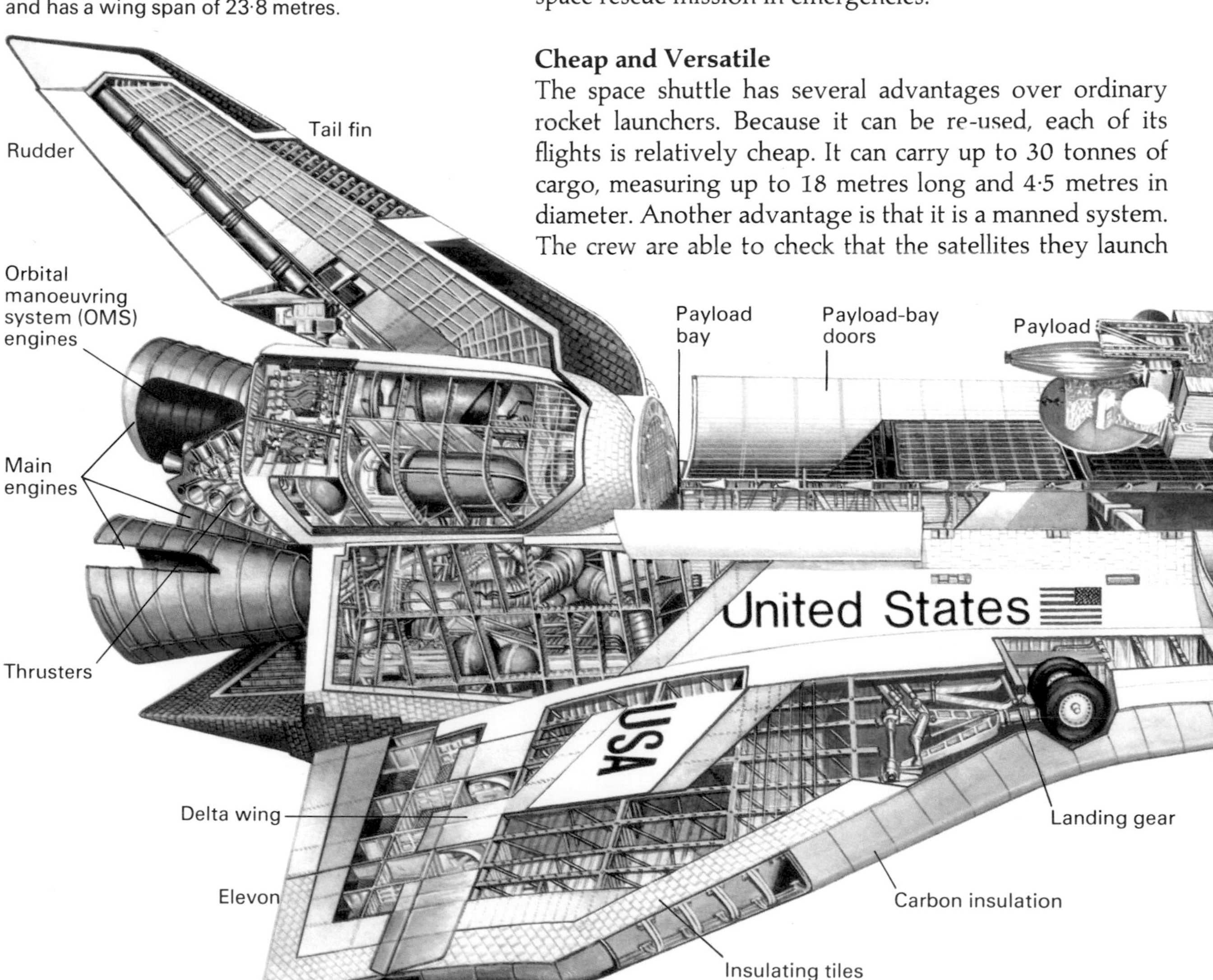

are working properly before they have to return to Earth.

The astronauts can also act as maintenance men and service satellites already in orbit. Or they can bring these satellites back to Earth. The space telescope (see page 167) is designed to be serviced in space and brought back to Earth from time to time.

Another special payload has been designed for the space shuttle. It is a space laboratory known as Spacelab, built by European space scientists (see page 178).

Countdown

Most space shuttle launches take place from the Kennedy Space Center at Cape Canaveral in Florida. There, the shuttle uses many of the facilities built originally for the Apollo Moon-landing missions.

The orbiter rides into space "piggy back" on a tank, which contains fuel for its engines. Attached to the sides of the tank are two more rockets, the boosters. Several weeks before launching, all these parts are put together in the huge Vehicle Assembly Building at the Kennedy Space Center (see page 176). They are mounted on a platform, which is then carried slowly to the launch pad on a massive crawler tractor.

A few days before launch the countdown begins. This is the time during which the shuttle is made ready for its flight. All its systems are checked to make sure they are working perfectly. Then the liquid hydrogen and liquid oxygen propellants are pumped into the fuel tank.

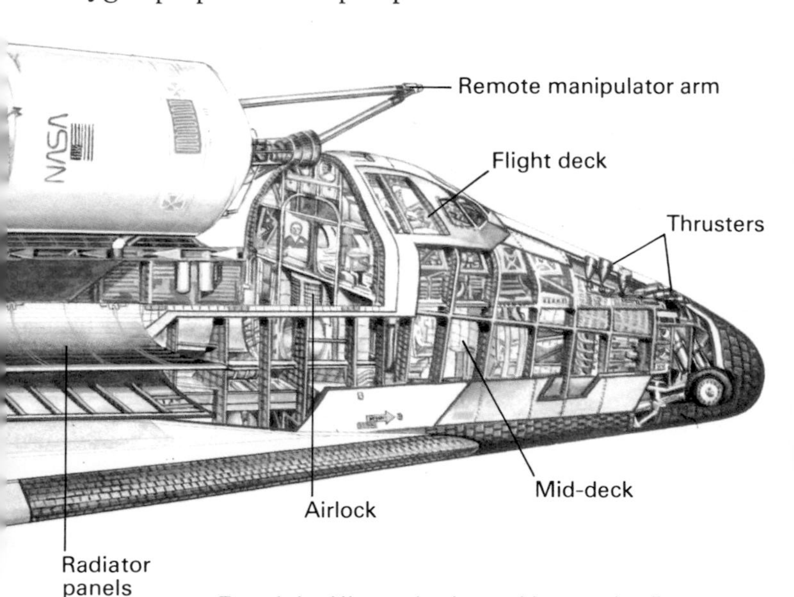

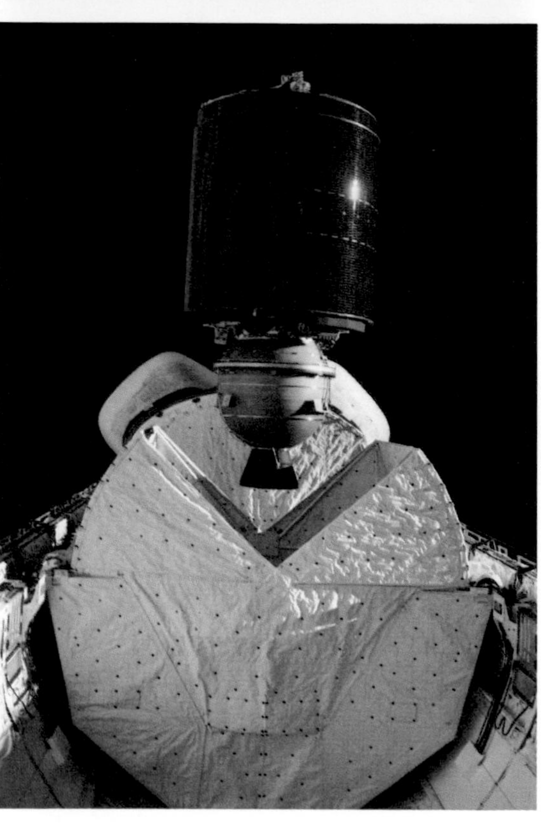

Top right: History in the making as the first space shuttle (STS-1) takes off from the Kennedy Space Center, on 12 April 1981. The orbiter is *Columbia*, which flew the first five shuttle missions.

Right: The Canadian communications satellite Anik C3 being launched from the payload bay of *Columbia* in November 1982 on the first operational flight of the shuttle.

Space Shuttle 2

Lift-Off!

At last everything is ready. The countdown clock ticks down to zero. The orbiter's engines and the rocket boosters fire, and the shuttle shoots off the launch pad on a pillar of flame and smoke. People watching from a safe 5 km away feel the ground shake and are deafened by the rockets' roar.

Two minutes after lift-off the boosters run out of fuel and fall away. They parachute down in the ocean from where they are recovered to be used again. Meanwhile the rest of the shuttle is speeding heavenwards, getting faster and faster. Every minute it burns 200,000 litres of fuel.

About eight minutes after lift-off the main engines shut down as the fuel tank runs dry. This separates in turn and smashes to pieces when it falls into the ocean. The orbiter has now reached a height of over 120 km and is still climbing. Two small engines then fire to boost its speed to 28,000 km per hour, and soon it is in orbit.

Return to Earth

After the crew have completed their work in space they prepare to return from orbit. First they turn round the orbiter so that its engines are pointing in the direction they

Right: The space shuttle orbiter *Columbia* a few seconds before touchdown at the Edwards Air Force Base in California. The orbiter takes off vertically like a rocket, but lands like a glider on an ordinary runway.

Bottom left: The shuttle hardware is fitted together in the huge Vehicle Assembly Building at the Kennedy Space Center. The orbiter is shown being lowered into position on its fuel tank, to which the two rocket boosters are already attached. The Vehicle Assembly Building was originally built to house the 111-metre tall Saturn V Moon rockets. One of the biggest buildings in the world, it measures some 218 metres long, 158 metres wide and 160 metres high.

Below: The various stages of a shuttle mission. Of the hardware that goes up, only the fuel tank is discarded. The solid rocket boosters and orbiter are re-usable. This is much more economical than using conventional launching rockets, which can be used only once.

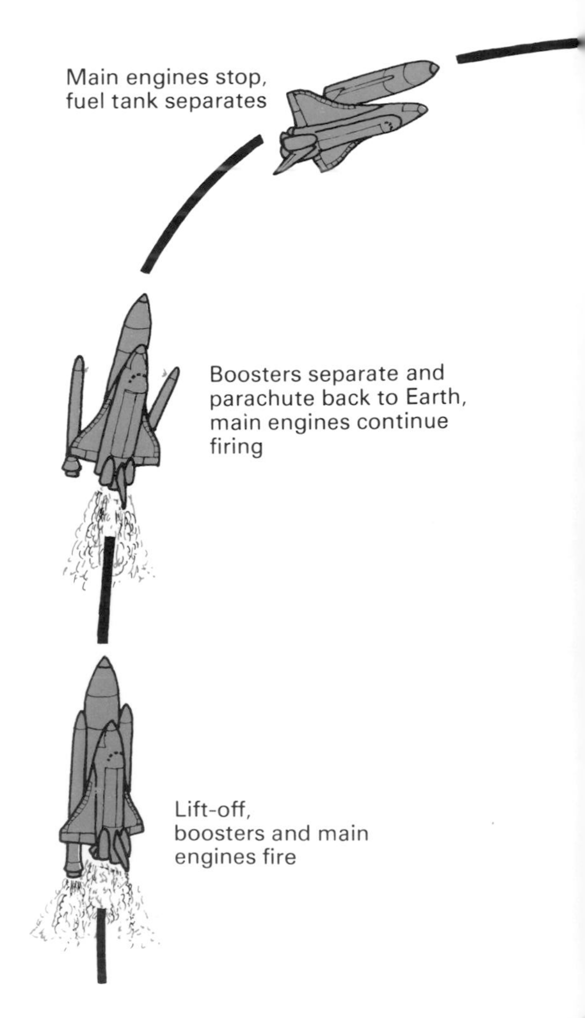

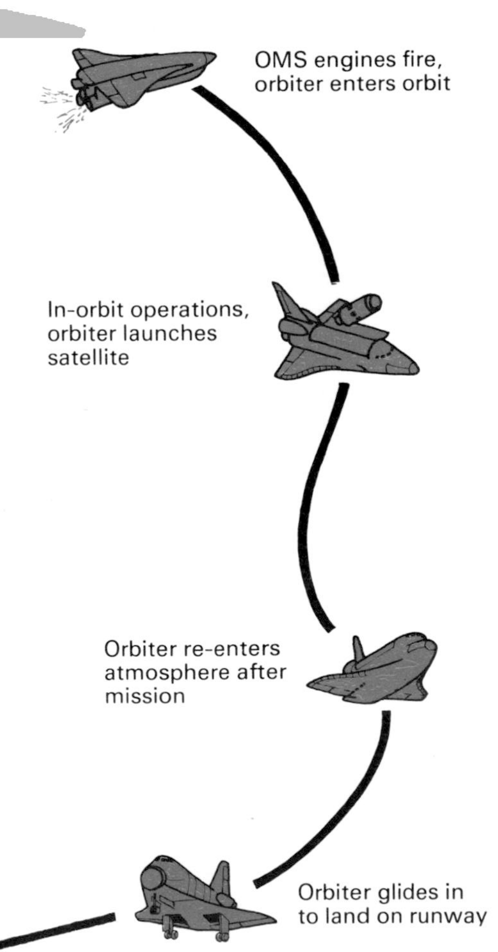

are travelling. They then fire the small engines again. This action, known as retrobraking, slows the orbiter down. And gravity begins to pull it back to Earth.

About 45 minutes later the orbiter re-enters the atmosphere. The air resistance, or drag, acts as a brake and rapidly slows down the craft. The outer surface heats up because of friction with the air. For protection the surface is covered with thousands of heat-resistant silica tiles.

Soon the orbiter is flying like a glider, using its wings and tail to manoeuvre in the air. (In space it had to use rocket engines to manoeuvre.) Finally it drops steeply down to land on an extra long runway. On touchdown the astronauts apply the wheel brakes, and the orbiter rolls to a halt. Although it takes off like a rocket, it lands like an aeroplane. It has stood up well to its journey into space. Within a fortnight it could be rocketing back there again.

The main shuttle landing site is also at the Kennedy Space Center, where the runway is nearly 5 km long. Early shuttle flights touched down at the Edwards Air Force Base in California, where there is plenty of room to overshoot the runway in safety. Edwards is one of several alternative landing sites for the shuttle scattered around the world which can be used in emergencies or when weather conditions at Kennedy are bad. There are others at Honolulu in Hawaii, Okinawa in Japan and Rota in Spain. Two other sites in the United States are at White Sands, New Mexico, and at Vandenberg Air Force Base in California, which is also a shuttle launch site.

Space Travellers 1

Above: The experimental American space station Skylab in orbit. Three teams of astronauts spent respectively 28, 59 and 84 days working in Skylab, setting new space endurance records.

Below: The Russian spacecraft Soyuz taking part in the joint American–Russian space mission, the Apollo–Soyuz Test Project, in 1975.

Unmanned satellites, although useful, are limited in what they can do. Manned spacecraft are much better because the astronauts on board can think for themselves and carry out all kinds of different activities. They survey the Earth and study the heavens. They investigate the effects of space on the human body and on plants and animals. They carry out experiments in all branches of science and engineering.

The manned spacecraft used by the Americans is the space shuttle (see page 174). Unlike previous craft it is reusable, being able to return to orbit again and again. The craft used by the Russians is Soyuz, which is an expendable craft – it can be used only once. Its main function is to ferry cosmonauts to and from the Salyut space stations.

Before 1961, no human being had ventured into space, although dogs and monkeys had been sent there and recovered successfully. So no one knew whether the human body could withstand the stresses and dangers of spaceflight. Then on 12 April 1961, the Russians launched the world's first spaceman, cosmonaut Yuri Gagarin. By 1969 three American astronauts were setting out in the Apollo 11 spacecraft on the most daring voyage of discovery in history. Their objective was to walk on the Moon.

On 20 July 1969, Apollo 11 astronaut Neil Armstrong planted the first human footprint on the Moon. He was the first of 12 American astronauts to explore the Moon on foot. The Apollo missions and the missions that followed, in experimental space stations such as Skylab and Salyut, proved beyond doubt that human beings can live quite happily in space. Only in the first few days of a mission are they usually affected – by nausea, or space sickness.

Space Stations

The Russians were the first to experiment with space stations, which are craft designed to remain in orbit for a long time. They launched their first craft, Salyut 1, in 1971, and in 1982 their seventh craft, Salyut 7. That year cosmonauts Anatoly Berezovoy and Valentin Lebedev in Salyut 7 set up a new spaceflight record by spending 211 days in orbit. During this time they were visited by a number of other crews in Soyuz ferry craft. From time to time they were brought fresh supplies by unmanned Progress craft, guided by remote control.

The first American experimental space station was called Skylab. It was made out of rocket parts left over from the Apollo Moon-landing programme. Three teams of astronauts visited Skylab in 1973/74. The last team spent 84 days there. Skylab was huge, measuring some 36 metres long and weighing 90 tonnes. The astronauts carried out many useful observations, particularly of the Sun.

Spacelab

Another important step in space-station design is Spacelab. This is a fully equipped space laboratory built by European space scientists, through the European Space Agency (ESA). Spacelab is designed to fit inside the payload bay of the space shuttle (see page 174). Like the shuttle it can be used again and again.

The main part of Spacelab is the laboratory module. This is pressurized with air so that scientists can work inside it without wearing spacesuits. The other part is an open platform (pallet) which only carries instruments. Up to four scientists or engineers can work in Spacelab. They share the crew quarters of the astronauts who pilot the shuttle. But they do not need to be fully qualified astronauts themselves. They can go into space after only a few weeks' special training.

In the future, units like Spacelab may be joined together to make much bigger space stations. They will each be ferried into orbit by a shuttle, and put together there. Construction engineers may use these stations as a base from which they can build even bigger space structures, which might include huge communications platforms or solar power satellites (see page 24).

Above: Astronauts can enjoy brief periods of weightlessness during training in a plane that flies up and over in a very tight arc.

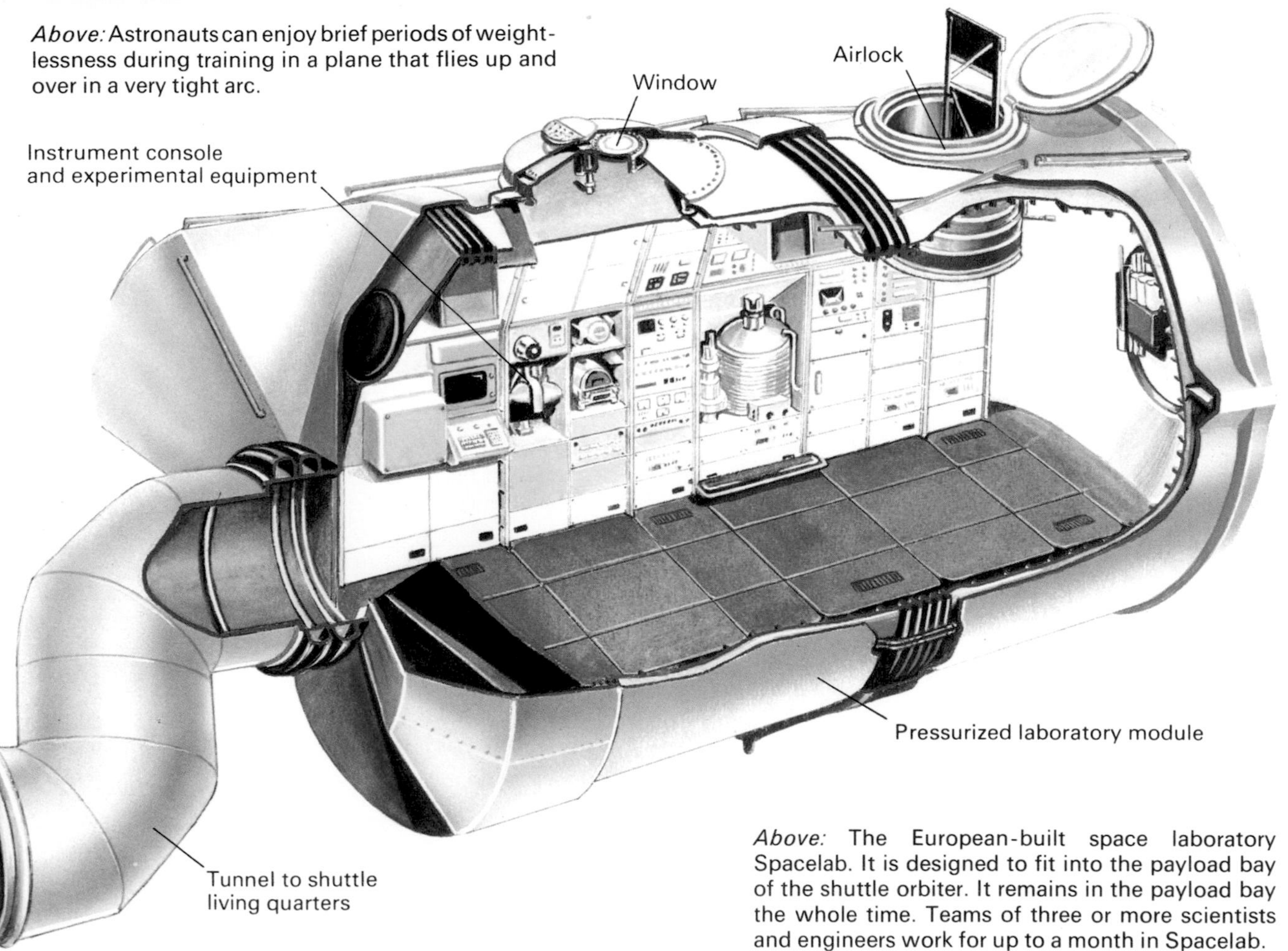

Above: The European-built space laboratory Spacelab. It is designed to fit into the payload bay of the shuttle orbiter. It remains in the payload bay the whole time. Teams of three or more scientists and engineers work for up to a month in Spacelab.

Space Travellers 2

Heavyweights

Imagine for a while that you are an astronaut about to travel into space. For the launching you are strapped into a seat with your back towards the rocket engines of your launching craft. When they fire, you are pressed against the seat with tremendous forces, called G(gravity)-forces. They make you more than three times heavier than usual (3Gs).

After about a quarter of an hour, the rocket engines shut off and you are in orbit 200 km above the Earth. Suddenly you realize a strange thing – instead of being three times heavier than usual, you are now weightless. You are actually "falling around the Earth" in a condition known as FREE FALL, or zero-G.

Weightlessness

Life is odd in this state of weightlessness. You can easily perform gymnastics, but you cannot walk properly for there is no normal gravity to keep your feet down on the floor. In fact there is no such thing as "down" or "up" in orbit. The only way to move is to pull or push yourself along. And the only way to stay still is to hang on to something, otherwise you just float about. The same problem occurs when you want to go to sleep. So you have to sleep zipped inside a sleeping bag that is anchored to something solid.

With no gravity to fight against in orbit, the muscles of your body tend to become flabby, rather as they do when you have to stay in bed for a long time when you are ill. So to keep fit, you must exercise regularly. For this reason spacecraft are fitted with exercise bicycles or treadmills. The Russians have developed special tight-fitting exercise suits that exert pressure on the muscles.

Zero-G also causes problems with drinking and eating. You cannot drink from a glass in orbit because the liquid

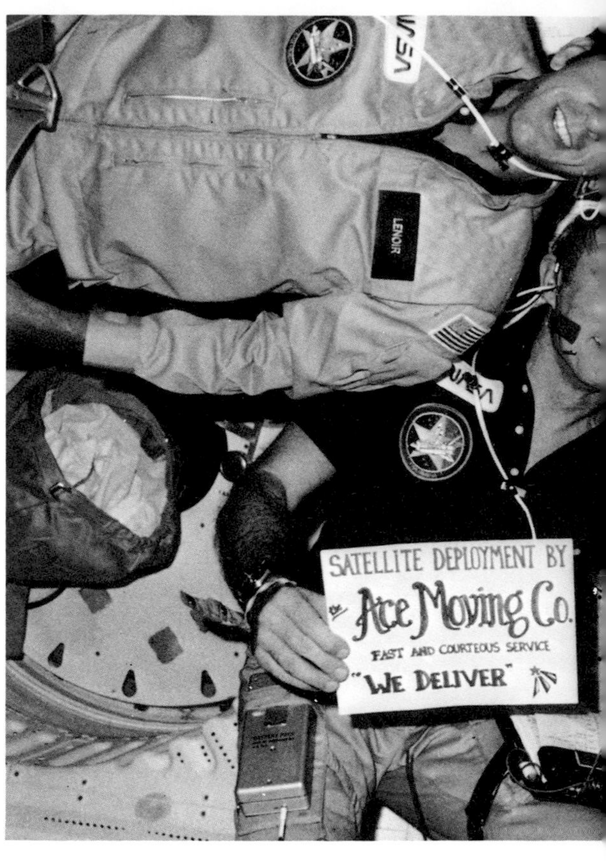

Above: Four astronauts in a weightless pose during the first operational flight of the shuttle in November 1982. They launched two satellites, which accounts for the message on the card.

Below left: American and European astronauts train in a mock-up of Spacelab.

Below: American astronaut Anna Fisher practises flight procedures in the space shuttle trainer. Women are playing an ever-increasing role in the American space programme.

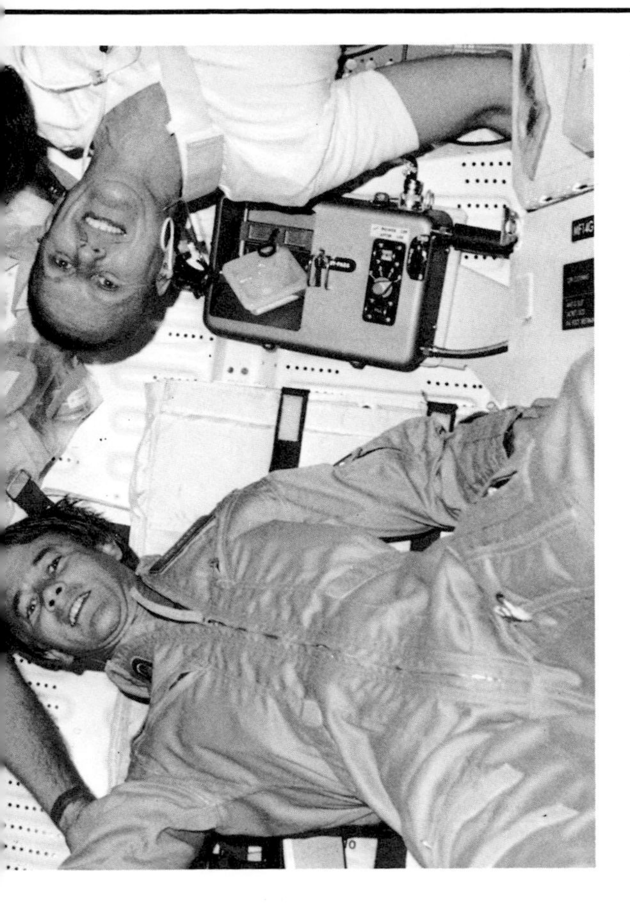

just stays in the glass. You have to drink by squirting the liquid into your mouth from a collapsible plastic bottle. You may also eat some foods directly from collapsible bags. But if you are on the space shuttle you will eat your main meals in more or less the normal way from containers on a tray, using knife, fork and spoon. The containers are held down on the tray by means of Velcro strips.

The foods you eat come from cans or flexible pouches. Many of them are supplied dehydrated, and you have to add hot or cold water to them before eating. It is necessary to eat moist foods which stick together, rather than crumbly foods, or else the crumbs will get everywhere. Bread, however, is still on the menu, but it is treated with gelatine so that it does not crumble so much.

Toilet Facilities

Toilet arrangements are also complicated by the state of weightlessness. Ordinary washing, for example, would result in showers of water droplets, which would spread everywhere. So in orbit you usually freshen up by rubbing yourself down with a moist cloth. On Salyut, where you might have to stay for months, you could enjoy the luxury of a shower inside a plastic closet. The closet is fitted with a suction device to remove the water.

Suction devices also feature in the lavatories now fitted to modern spacecraft, in which air not water is used for "flushing". In the space shuttle lavatory, for example, urine is removed through a hose and stored in a tank for disposal back on Earth. Solid wastes are sucked into another tank from a commode and later dried by being exposed to the vacuum of space. They are also returned to Earth.

Below: The manned manoeuvring unit developed to assist shuttle astronauts working in orbit.

Life-Support

Space is a really deadly place for human beings. There is no air to breathe. Temperatures are scorching hot in the Sun and freezing cold in the shade. Dangerous particles stream from the Sun, and tiny specks of rock form a ceaseless bombardment. To survive, human beings must be protected from all these things.

This is the job of their spacecraft's life-support system. It provides astronauts with oxygen to breathe and removes stale air and odours. It keeps the temperature and humidity of their spacecraft at comfortable levels. It is similar to the pressurized air-conditioning systems that airliners have.

Sometimes astronauts have to work outside their craft, on what is called extra-vehicular activity (EVA). Then they wear a spacesuit, which is made up of several garments worn one on top of another. The inner one is water-cooled to control the astronaut's temperature. The suit may be linked by a tube to the spacecraft's life-support system. Or it may be supplied from a backpack.

Space Probes

Satellites and space stations remain quite close to the Earth, locked in orbit by gravity. Other spacecraft, however escape from gravity completely and voyage into the depths of space. They are called space probes. Some of the most exciting discoveries in astronomy in recent years have been made by space probes. By 1979 the five nearest planets had been explored by probes such as Mariner, Viking, Voyager and Venera. In 1983 the Jupiter probe, Pioneer 10, left the solar system and headed into interstellar space.

Mariner 10 showed that the planet Mercury looks very much like the Moon. The Russian Venera probes reported that Venus is hotter than molten lead. Viking showed massive volcanoes and huge canyons on Mars. Voyager spotted volcanic eruptions on one of Jupiter's moons, and revealed that winds on Saturn blow at speeds up to 1800 km an hour. The Venera and Viking probes actually landed on the planets they explored.

Escaping from the Earth

It is very much more difficult to launch a probe to the planets than it is to launch a satellite into orbit. A probe has to escape completely from the powerful pull of Earth's gravity. It can only do so if we give it the colossal speed of over 40,000 km an hour. This speed is called the escape velocity.

Another great problem with probes is aiming them accurately. Remember that their target – a planet – is many millions of kilometres away, and is moving. The probe must be aimed so that it reaches a certain point in space at the same time as the planet. The problem becomes greater the farther the probe has to go and the longer its journey takes. Yet, thanks to powerful computers, its path through space can be worked out and its time of arrival forecast accurately.

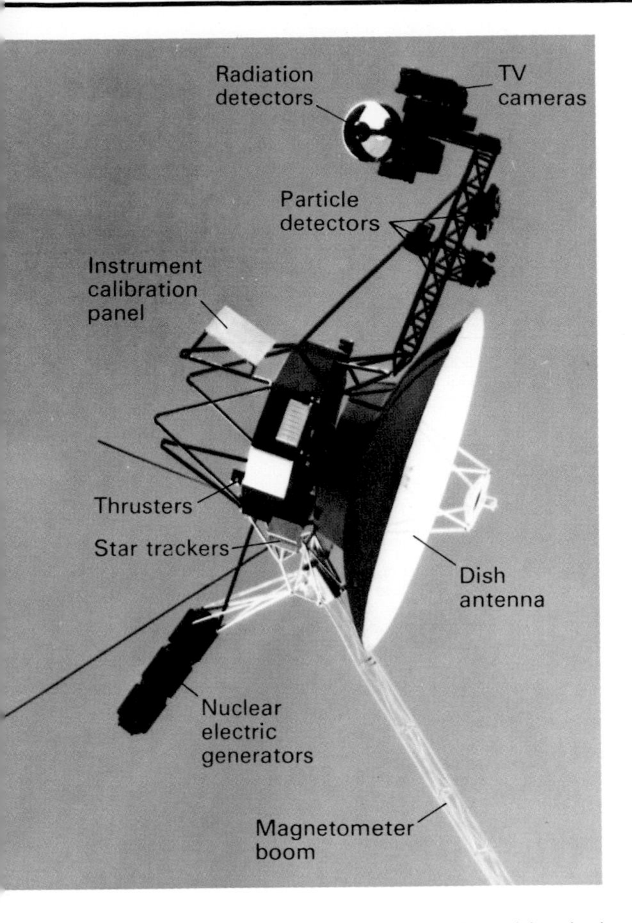

Above: The Voyager space probe. Two identical Voyager probes set out in 1977 to explore the outer solar system and sent back the most spectacular pictures of Jupiter and Saturn. Voyager 2 is now heading for Uranus.

Top left: The beautiful ringed planet Saturn, pictured in false colours. The image was produced from data sent back by the Voyager 2 space probe in 1981. The false colours show up more detail than the true colour can. The parallel bands on the planet's surface are high-speed wind streams.

Far left: The Viking lander, two models of which landed on Mars in 1976. The landers took close-up pictures of the Martian surface and reported on Martian weather. They also searched for life, by analysing the Martian soil in automatic laboratories. But they searched in vain.

Left: The surface of Mars, as pictured by the Viking 1 lander, which set down in a region called Chryse. The soil is rusty, reddish-brown in colour and is strewn with many small rocks. It gives the whole planet a reddish tinge, which we can see from Earth. Astronomers often call Mars the Red Planet.

Right: The European space probe Giotto, designed to rendezvous with Halley's comet when it comes close to Earth in 1986.

Communications

Another astonishing thing is how scientists on Earth manage to communicate with a tiny probe far away in the depths of space. Again, computers help them pinpoint the probe's exact position so they can point their radio aerials in the right direction. These aerials have to be huge: the Goldstone antenna in California is a dish 64 metres across. It has to be as big as this otherwise it could not detect the faint signals sent out by a probe. These signals may take hours to travel to the Earth.

Voyager's Voyage

For the past few years the American Voyager 2 probe has been carrying out one of the most demanding missions of the space age. It was launched in August 1977. After a journey of about 1000 million km, it reached Jupiter in July 1979 and took thousands of beautiful pictures of that giant planet and its moons. Then it headed towards Saturn, where it arrived in August 1981 after travelling another 1000 million km.

Leaving Saturn and its spectacular rings behind, Voyager then began its 1000-million-km journey to the planet Uranus. That is where it is heading now. All being well, it should pass close to Uranus on 24 January 1986. Even then its journey will not be over. The next stop will be the planet Neptune in August 1989.

Eventually Voyager will escape from the solar system and head for the stars. Many thousands of years hence it might reach another planet in another solar system. If intelligent beings live on the planet and find Voyager, they may play the record it carries. This record, called "Sounds of Earth", contains greeting from Earth people in 60 languages and sounds of the natural and man-made world. So perhaps one day other beings in the universe may know about us and our 20th century civilization.

Glossary

AEROFOIL The shape of a cross-section of an aeroplane's wing. It is broad and rounded in front and pointed at the rear. The top is curved while the bottom is relatively flat. When air moves over such a shape, the pressure underneath is higher than that on top, and this creates "lift".

ANALOG COMPUTER A type of computer that handles data in the form of physical values such as temperature, voltage and pressure, rather than in the form of coded digits.

ANTIBIOTICS Chemical compounds produced by microorganisms, such as bacteria and moulds, which are used as drugs to combat disease.

ARTIFICIAL INSEMINATION A technique used to improve the quality of livestock breeds, in which semen is introduced artificially into female animals.

ATOMIC BOMB Or A-bomb; a bomb that uses the enormous energy released during nuclear fission. The first A-bomb was detonated in New Mexico, USA, in July 1945. The following month the Japanese cities of Hiroshima and Nagasaki were bombed, ending World War 2. They each had the explosive force of some 20,000 tonnes of TNT, and together killed over 100,000 people.

AUTOMATIC PILOT A gyroscopic device, often nicknamed "George", that pilots a plane automatically by maintaining a pre-set course, altitude and speed.

BALANCE A sensitive weighing machine used in scientific laboratories. The traditional type uses the balancing principle of the lever to determine weight; the scale pans hang from an arm that pivots about a knife edge.

BINARY NUMBER A number system that has a base of 2 and uses only the digits 0 and 1. Place values in the system go up in powers of 2 (instead of powers of 10 as they do in our normal decimal, or base 10, number system). In binary, the decimal number 25 is represented as 11001 ($1 \times 2^4 + 1 \times 2^3 + 0 \times 2^2 + 0 \times 2^1 + 1 \times 2^0$).

BINOCULARS A compact type of telescope that has two tubes, one for each eye. The instrument is compact because it incorporates pairs of prisms to "fold" the light path between the eyepiece and objective lenses.

BLACK HOLES Awesome bodies that astronomers believe exist in the centre of galaxies. They have such an enormous gravitational pull that nothing, not even light, can escape from them.

BOTTLED GAS A useful portable fuel, consisting of liquefied petroleum gases such as butane or propane. They are kept liquid under pressure in strong containers. The butane or propane turns into gas again when the pressure is released.

BUBBLE CHAMBER A device used in nuclear physics to detect and show the tracks of atomic particles. When particles pass through the chamber, which is usually filled with liquid hydrogen, they leave a stream of bubbles in their wake, which can be photographed.

CAISSONS Cylindrical or box-like structures that are sunk to the river bed to provide the foundations for a bridge. Then they are filled with concrete.

CARBON FIBRE Threads of carbon that are used as reinforcement in plastics, ceramics and metals. They are light and exceptionally stiff and have many uses, particularly in aircraft construction.

CATALYSTS Substances that help a chemical reaction take place without themselves being changed. They are widely used in the chemical industry. Platinum is a good catalyst, as are iron and nickel.

CATHODE-RAY TUBE (CRT) The picture tube used in television receivers, oscilloscopes and similar instruments. It is an evacuated glass tube, narrow at one end and broad at the other. At the narrow end is an electron gun which "shoots" out a beam of electrons (or cathode-rays). At the broad end is a fluorescent screen, which glows when the electrons strike it. In between there are a number of coils which make the electron beam trace patterns—pictures or signals—on the screen.

CERMET A strong heat-resistant material made from a combination of a ceramic and a metal. Examples are combinations of titanium carbide and cobalt; aluminium oxide and chromium; and chromium carbide and nickel.

COAL GAS The gas produced when coal is destructively distilled, or heated in the absence of air. The main gases produced are methane, hydrogen and carbon monoxide.

COAL TAR A tarry mixture of substances obtained from the vapours given off when coal is destructively distilled (see above). It is a rich source of organic chemicals.

CONCRETE Our most useful building material, made from a mixture of cement, gravel, sand and water.

CONVEYOR A device for moving things. It often takes the form of a moving belt, a rotating screw or a moving chain.

CRYOGENICS The study and application of very low temperatures. It is particularly concerned with the use of liquid gases such as liquid oxygen (boiling point, $-183\,^{\circ}C$) and liquid helium ($-269\,^{\circ}C$).

DISTILLATION A common process used in the laboratory and the chemical industry for purifying and separating liquids. It consists of heating a liquid to boiling point and then condensing the vapour given off.

DNA An organic compound that is the "brains" behind the workings of the living cell. Its chemical name is deoxyribonucleic acid.

ECG An abbreviation for "electrocardiograph", an instrument that records the electrical activity of the heart. From an ECG trace doctors may be able to diagnose heart complaints.

ECHO-SOUNDER A device that uses sound echoes to determine the depth of water beneath a vessel. It works by sonar. It sends out high-pitched, ultrasonic pulses, and measures the time it takes for them to be reflected from the seabed. From this, it works out the depth.

EEG An abbreviation for "electroencephalograph", an instrument that records the minute electrical activity of the brain. From an EEG trace doctors may be able to diagnose brain abnormalities.

ELECTROLYSIS The splitting up of a chemical compound by means of electricity. When electricity is passed through an electrolyte such as a salt, in solution or when molten, the salt is split up into its constituent elements.

ELECTRONIC FLASH A device used in photography that provides a brief, very intense light. It consists of a tube containing a gas such as krypton, and gives out a bright flash when a high-voltage pulse is passed through it.

ELECTRONICS A branch of science and technology concerned with devices controlling the flow of electrons, such as vacuum tubes, transistors, silicon chips and similar semiconductor devices.

ELEMENTARY PARTICLES Particles found in, or emitted from the atom. They are also called subatomic and fundamental particles. The most important ones are protons, neutrons and electrons, found in all atoms except hydrogen (which lacks neutrons). Other particles include positrons (positively charged electrons), mesons, neutrinos, and quarks.

ENDOSCOPE A fibre-optic device used to see inside things, including the human body. It is a narrow, flexible rod made up of bundles of glass fibres, which can be inserted down into the stomach for example. Light shines down some of the fibres, while others are used for viewing.

ENZYMES Substances found in living things, which act as catalysts to bring about chemical reactions, such as digestion. They are kinds of proteins. In industry enzymes are used to assist fermentation processes in brewing for example; in food processing and in detergent manufacture.

EUTROPHICATION A form of pollution caused by the run-off of agricultural fertilizers into rivers and lakes. The fertilizers promote the growth of water plants, which extract all the oxygen from the water, eventually causing the death of water life.

FACSIMILE Or fax; the sending of pictures, plans or documents by wire. The information in a picture, for example, is converted by an optical scanning method into electrical signals, which are then transmitted. In the receiver the electrical signals are converted back into a visible picture.

FIBRE OPTICS A developing field of technology in which bundles of fine glass fibres are used to transmit light. The endoscope is a fibre-optic device. Fibre-optic cables are now coming into use in telecommunications networks in place of copper cables and carry signals as coded pulses of laser light.

FREE FALL A condition that exists in orbit when bodies are travelling around the Earth. They are actually falling towards the Earth, but the amount they fall is the same as the amount the Earth curves away. So they in effect remain at exactly the same height. This condition is commonly called weightlessness.

FUEL CELL A device that produces electricity by "burning" fuels chemically by means of a catalyst. The commonest type of cell uses hydrogen and oxygen, which are made to combine to form water. The space shuttle uses this kind of fuel cell.

GAS CHROMATOGRAPHY A very sensitive method of separating complex substances such as a mixture of organic chemicals. The mixture is vaporized and carried by an inert gas such as nitrogen through a column packed with adsorptive material (one that attracts substances to its surface). The chemicals in the vapour take varying times to pass through the column and are thereby separated.

GAS TURBINE An engine that burns fuel in compressed air to form hot gases that spin a turbine. In a typical gas turbine, air is taken in and compressed by a compressor. Fuel is then mixed with the compressed air in a combustion chamber and then burned. The hot gases produced spin one or more turbines. One turbine at least drives the compressor.

GEIGER COUNTER A common instrument for detecting and measuring atomic radiation. It consists of a gas-filled tube charged to a high voltage. The passage of an atomic particle through it causes an electric discharge. This triggers a counting device.

GENETIC ENGINEERING Changing the genetic make-up of living things. Scientists are now able to isolate the genes in the DNA of certain organisms and transfer them to the DNA of other organisms, making these behave differently. So far this technique has been carried out successfully only in micro-organisms such as bacteria.

HOLOGRAPHY A technique of three-dimensional (3D) photography using a laser beam. In holography a laser beam is split into two. One part is shone onto photographic film directly. The other is shone on the object to be photographed, and light is then reflected onto the film. When laser light is shone through the developed film ("hologram"), a true 3D image appears.

HORSEPOWER A common English unit of power, introduced by James Watt as the power of an average working horse. One horsepower is equivalent to 746 watts, the power unit in the metric system. The horsepower usually quoted for car engines is the so-called brake horsepower. This is a measure of the actual, rather than the theoretical power output of the engine.

HYDROCARBONS Organic substances made up of the elements hydrogen and carbon only.

HYDROGEN BOMB Or H-bomb; the most deadly of weapons, which uses the principle of nuclear fusion. In this process atoms of heavy hydrogen are made to combine, or fuse, into atoms of helium, and enormous energy is released. H-bombs have an explosive power equivalent to millions of tonnes ("megatons") of conventional explosive such as TNT. They use an atomic bomb as a trigger to produce the high-temperatures necessary for fusion to occur.

HYDROPONICS The practice of cultivating plants without soil. They are grown instead in solutions containing just the right nutrients and chemicals necessary for maximum growth.

LCD An abbreviation for "liquid-crystal display", the type of display widely used in digital watches and pocket calculators. The display contains a thin layer of a liquid that behaves like a crystal and can twist the path of light passing through it. The crystal untwists when electricity is applied to it. This phenomenon is used to prevent the reflection of light from segments of the display, which thus appear dark, and these form the digits.

LED An abbreviation for "light-emitting diode", one way of forming the digits in the display of a calculator, clock, or other device. The digits are formed from patterns of tiny crystal diodes, which emit light (usually red) when electricity is passed through them.

LOUDSPEAKER A device that changes the electrical signals produced in a radio, record player, and so on, into sound. The signals are passed through a fine coil of wire, located inside a magnet, and cause it to vibrate. The vibrations are passed to a paper cone, which sets up the sound waves. Hi-fi systems generally have two or more speakers, each reproducing sound of a particular frequency range.

MACH NO The speed of an aircraft expressed relative to the local speed of sound. Mach 1 equals the speed of sound; Mach 2 equals twice the speed of sound. It is named after the Austrian physicist Ernst Mach.

MARGARINE A butter substitute made from various vegetable oils, animal fats and milk products. The original margarine was produced by a French chemist, H. Mège-Mouries, in 1869.

MICROELECTRONICS The branch of electronics that deals with the production of microcircuits—microscopic electric circuits formed in silicon chips.

MICROFILM Pieces of photographic film containing greatly reduced images of newspapers, documents, and so on. For a typical application—say, the storage of newspapers—the reduction may be about 20:1. The film used may be 16 mm, 35 mm or 105 mm wide.

MICROMETER GAUGE Also called a micrometer screw gauge; a G-shaped instrument widely used in engineering and industry to measure dimensions. It uses the property of a screw that when it is rotated once, it advances a precise amount, equal to the pitch of its threads. The latest micrometers show measurements on a digital display.

MICROPHONE A device that converts sound waves into electrical signals for transmission by wire or radio. Each type of microphone has a diaphragm, which the sound waves vibrate. The vibrations are then converted into electrical signals in various ways, for example, by means of a moving coil or a piezoelectric crystal.

MICROWAVES Very short radio waves used in radar, for transmitting signals in communications, and in the home for microwave cooking. They have a wavelength from about 1 millimetre to 30 centimetres.

MOTION PICTURES Films shown on a screen that give the impression of continuous motion. The movement is actually an optical illusion. What happens is that a series of still photographs, taken in rapid succession (24 pictures, or frames per second), is projected onto the screen at the same rate. Due to a phenomenon called the persistence of vision, the eyes hold onto an image of one frame for a fraction of a second while the next frame is projected. The images merge and the impression is given of continuous motion.

MUTATIONS Alterations that occur in the genetic make-up of living things, which cause them to change in one way or another. Chance mutations in genes have occurred since life on Earth began, giving rise to the evolution of new life forms. Radiation is one agent that can bring about mutations. Scientists today use radiation to try to cause mutations in plant species that may result in improved varieties. The prospect of harmful mutations is one danger associated with exposure to nuclear radiation.

NUCLEAR FISSION The splitting of the nucleus of a heavy atom, particularly uranium, a process accompanied by the release of enormous energy. Uranium atoms are unstable and sometimes split into smaller atoms. They can also be made to split by bombarding them with neutrons. When they split, they give off two or more neutrons. These neutrons can go on to split other uranium atoms; and the neutrons given off in these fissions can in turn split still more uranium atoms. Such a "chain reaction" results in the release of fantastic amounts of energy. This energy release is controlled in nuclear reactors; uncontrolled in the atomic bomb.

NUCLEAR FUSION The joining together of the nuclei of light atoms to form a heavier one, a process accompanied by the release of large amounts of energy. Fusion reactions in the Sun and the stars provide the energy to keep these bodies shining. They involve the fusion of hydrogen into helium. Scientists have imitated the process on Earth, using heavy forms of hydrogen called deuterium and tritium. The result was the hydrogen bomb. They are now trying to harness the fusion process for power generation.

NYLON A widely used synthetic fibre and plastic, first produced in the early 1930s by the American chemist W. H. Carothers. It is classed as a polyamide.

OSCILLOSCOPE A device that uses a cathode-ray tube to display the relationship between variable electrical quantities, such as voltage and current.

OTEC An abbreviation for "ocean thermal energy conversion". This is a scheme proposed for extracting the energy from hot tropical waters.

PASTEURIZATION A method of temporarily sterilizing milk by heating it for about 15 seconds to a temperature of about 70°C. It is named after the French chemist, Louis Pasteur.

PEAT A low-grade fuel made up of partly decayed plant matter. Eventually, given the right geological conditions, peat could turn into coal.

PHOTOCOPYING A method of copying papers, drawings, and so on. A variety of photocopiers are in use. The simplest use sensitive paper "negatives", and chemicals or heat are used to develop an image. The most advanced machines use the technique known as xerography, which works by the transfer of electric charges. In the process an electrically charged image is formed on a rotating drum and attracts ink powder. The ink image is then transferred to paper and fused to it by heating.

POLYGRAPH The correct name for the instrument known popularly as a lie detector. It measures various body functions of the person being interrogated, such as pulse rate, respiration and blood pressure. When the person lies, the instrument should register slightly different readings.

PRODUCER GAS A low-grade fuel gas used in industry, made by heating coal with air and steam. It contains carbon monoxide and hydrogen.

PULSARS Small heavenly bodies that rotate rapidly and give out regular pulses of radio waves and often other radiation as well. They are believed to be made up of solid neutrons and are of such high density that a tablespoonful would weigh 1000 million tonnes!

QUASARS Or quasi-stellar objects, mysterious heavenly bodies that appear like ordinary stars through a telescope, but are much farther away. They are very much smaller than galaxies, yet are many times brighter than galaxies. And they appear to lie thousands of millions of light-years away, at the very edge of the observable universe.

RADAR A major aid to aerial and marine navigation, developed in Britain in the late 1930s. It works by the transmission of microwave pulses and the detection of their echoes. The word is an abbreviation for "radio detection and ranging". The microwave pulses are emitted by an aerial, which also receives any echoes. From the time between transmission of a pulse and reception of its echo, the distance to the object causing the echo is determined. The result is displayed as a spot, or "blip" on a radar screen, which is a cathode-ray tube similar to that in a television receiver.

RADIOACTIVE Giving off radiation. Most of the chemical elements are stable, remaining the same all the time. Some, however, are unstable, and may suddenly break down, giving off radiation and atomic particles. This phenomenon is called radioactivity. The original element changes into another, whose atoms are different.

REACTION PRINCIPLE The principle that explains why jet and rocket engines work. It was first stated by Isaac Newton in the 1600s as his third law of motion: to every action there is an equal and opposite reaction. In the case of the engines mentioned above the force of the jet of gases escaping backwards is accompanied by an equal and opposite force acting forwards, which provides the propulsive thrust.

RESISTOR A device used in electrical circuits, which provides a certain resistance to the flow of electricity.

SCP An abbreviation for "single-cell protein". It is an artificial foodstuff made by processing simple organisms such as algae, yeasts and bacteria.

SEISMOGRAPH An instrument that records the ground vibrations (seismic shocks) set up by an earthquake. One type consists of a heavy weight suspended in a frame by a spring. When the ground moves suddenly, the frame also moves, but the weight, because of its inertia, stays where it is. The relative movement between frame and weight is fed to a recording pen.

SEMICONDUCTOR Material that conducts electricity slightly under suitable conditions. The commonest semiconductor is silicon, which is the basis of many modern electronic devices such as transistors and silicon chips. The silicon is made into a conductor by introducing small amounts of impurities. Other semiconductors include germanium, selenium and tellurium.

SILAGE A nutritious animal feed made by fermenting green grass and other crops, such as lucerne.

SI SYSTEM The international system of units of measurements now used extensively in science and technology. It is based on the fundamental units of metre, kilogram, second, ampere and kelvin (for temperature).

SOIL MECHANICS A branch of construction engineering dealing with the properties of soils, with particular regard to their ability to bear loads.

SONAR An invaluable aid to navigation at sea, which uses ultrasonic waves. It is a kind of sound radar. a pulse of ultrasonic waves is transmitted, and a receiver picks up any echoes reflected from objects in the surrounding water. These can then be located. Submarine craft use sonar for navigation underwater. Surface vessels use it, in the echo-sounder, to find the depth of water beneath them. Sonar techniques are now also used in medicine to look inside the body.

SOUND BARRIER A term once widely used in aviation, when it was thought that no aircraft could go faster than the speed of sound (about 1220 km per hour at sea level). At this speed shock waves buffet an aircraft.

SPECTROSCOPE An instrument for splitting light or other radiation into a spectrum. In the common light spectroscope, the light passes through a narrow slit and is formed into a parallel beam by a lens (collimator). The beam then passes through a prism (or a diffraction grating) and fans out into a spectrum, which is then viewed through a telescope or photographed.

STEREOPHONIC A sound reproduction system that produces "left" and "right" sounds. When we listen to an orchestra, say, live, slightly different sounds reach our two ears and give the music depth. Stereophonic systems try to imitate this process.

STIRLING ENGINE A piston engine that works on hot air, invented by a Scottish clergyman, Robert Stirling, in 1876. It works efficiently and causes little pollution.

STROBOSCOPE An instrument that employs a rapidly flashing light to measure the speed of rotating machinery. By shining the light at a rotating shaft, and adjusting the rate of flashing, the shaft can be made to appear to stop.

SUEZ CANAL The first of the great sea canals, built by the French diplomat and engineer F. de Lesseps and completed in 1869. It runs for some 160 km from Suez to Port Said.

SUPERCONDUCTOR A metal that completely loses its electrical resistance when it is cooled to within a few degrees of absolute zero (−273°C). Tin, lead and niobium are examples of superconductors. Very powerful electromagnets can be made using coils of superconductors.

SUPERNOVAE Stars late in their life cycle that suddenly explode and blast most of their matter into space. Only three supernovae have been seen in our galaxy in the past 1000 years—in 1054, 1572 and 1604.

SYNTHETIC RUBBER A kind of elastic plastic that resembles natural rubber. The first successful synthetic rubber, neoprene, was produced in the 1920s by W. H. Carothers in the United States. One of the commonest synthetic rubbers used today is styrene-butadiene rubber (SBR).

TRANSFORMER An electrical device for increasing ("stepping-up") or decreasing ("stepping-down") the voltage of alternating-current electricity. A simple transformer consists of two coils wound around a common iron core. When low-voltage electricity is applied to one coil, which has few turns, it sets up high voltage in the other coil, which has many turns.

TRANSISTOR A component used in radios and other electronic devices, made usually of silicon crystals. It is used to amplify or control the flow of electrons in a circuit, behaving like an old-fashioned radio valve. But it is much smaller, tougher and requires less power.

TVP An abbreviation for "textured vegetable protein", a manufactured foodstuff often made from soya beans. Protein is extracted from the beans and then dissolved in alkali. The solution is then extruded through a spinneret into an acid bath. The fibres that form are gathered into a rope, which is then chopped up and blended with fat, flavouring and other additives.

WANKEL ENGINE A novel rotary petrol engine, invented in the 1950s by the German engineer Felix Wankel. It uses a triangular rotor, rotating inside a figure-of-eight shaped chamber. It is much smoother in operation than the normal petrol engine, but various design problems have so far limited its use.

WIND TUNNEL A device used as an aid to aircraft design. In a wind tunnel, air is blown past a scale model of an aeroplane, and the behaviour of the model is noted. From this the designers can gain a very good idea of how the real plane will behave when flying. Designers of cars and bridges are among engineers in other fields who now make use of wind tunnels in their work.

Index

(Figures in **bold** refer to section headings; figures in *italics* refer to illustrations)

Acknowledgements

The author would like to extend his grateful thanks to the many organizations and individuals at home and abroad, who unstintingly provided information and pictures for the book. He is particularly indebted to Ian McIntosh and John and Mike Gilkes who prepared most of the artwork. He would also like to thank Gael Hayter and Penny Warn for assistance with the picture research. The credits for individual pictures appear below. Pictures not credited were taken by the author.

PICTURE CREDITS (T = Top, B = Bottom, C = Centre, R = Right, L = Left)

Aerofilms 86
Aérospatiale 110BL
Airbus Industrie 110BR, 113
Atlas Copco 83T, 84T
Audi-Volkswagen 91
Austin-Rover 64T
Australian News & Information Bureau 42, 172B
Australian Overseas Telecommunications Commission/ESOC 138
Bayer Agrochemicals 60
BP photographs 10T, 12, 14, 51, 57, 58, 78B, 82, 84B, 115T, 128B, 134T, 146T, 157T
British Aerospace 114C
British Airways 108T, 111BR
British Hovercraft Corporation 120B, 121TR
British Oxygen Company 48T
British Rail 102B
British Telecom 132B, 137
California Institute of Technology 169BR
Canon 125B, 145T, 147T, 150
Casio Electronics 133T, 144C
CEGB 16, 21
CERN 147BR, 153, 158B
Cerro Tololo Inter-American Observatory 164B, 165B
Ciba-Geigy 61B
R. O. Coffin 100BL
Commodore 147BL, 151T
Costain PLC Group 76B, 80T, 106
Courtaulds/Barry Finch Photography 52T
CPT 120BL
Culham Laboratory 25B
Daily Telegraph 67T
Deutsche Bundesbahn 102T
DJB Engineering 72B, 96C
Eastern Daily Press/RSPCA 69B
Harold Edgerton/Science Photo Library 131T
Electrowatt Engineering Services 83BR
EMI Medical 159T
Eurocom 129B
European Space Agency 170B, 173, 183B
Feedex Pig Partnership 33T
Ferranti Electronics 64B
Fiat 34, 50T, 66T, 103B
Ford 38, 90
Fujitsu Fanuc 65B
Fullwood & Bland 149T
Hall Automation 66B
Hewlett Packard 54T, 148B
Hong Kong Mass Transit Railway 104T
IBM 144T
ICI Plastics 62T
Inmarsat 116B
Intersub 141B
Japan National Railways 100T, 103T
JCB 73B
Jet Propulsion Laboratory 169TL, 182R, 183T
Kitt Peak National Observatory/Spacecharts 155B, 164T, 166T, 169TR, 169BL
Kodak 128T, 129T
Langley Research Center 130T
Lawrence Livermore Laboratory 154T, 155T
Leverton/Caterpillar 72T
Lockheed 54B
Marconi 30B, 140T, 141T
Martin-Marianetti 182BL
Massey-Ferguson 26, 28B, 32B
Medata Systems 28T
Metro-Cammel Weymann 96T, 99T
Microcomputer Printout 148T
Milk Marketing Board 37T
John Mills Photography 70, 83BL
Jean-Pierre Moutin/Frank Spooner Pictures 69T
NASA/Spacecharts 24T, 43B, 139T, 140B, 168B, 170T, 171, 174T, 175B, 176–181
NCB 8
Norwegian Caribbean Lines 118T
Novosti 121TL, 172L
Nuclear Engineering International 19
OCL 98T, 116T, 119
Parkes Radio Observatory 162
Robert Pendreigh/Royal Hampshire County Hospital 160T, 161B
Pest Information Control Laboratory/Crown Copyright 157B
Philips 146B
Pilkingtons 55T
Platt Saco Lowell 52B
P & O Lines 120T
Reed Paper & Board 29T, 41, 124T
Renault 92, 94
Rolls-Royce 45
Royal Astronomical Society 168T
Ruston-Bucyrus 40
Sandia Laboratories 25T
Clive Sawyer/ZEFA 74BR
Sea Fish Industry Authority 30T
Sharp Electronics 144B
Sheffield Forgemasters 46R, 49, 50B
Shell Photographs 15, 59B, 88
Smiths Industries 151B
SNCF 100BR
Snowy Mountains Authority 20
Société Bertin 105B
Sony 129C, 135BR
Spacecharts 174B
Sperry New Holland 32T, 35B
Stanford Linear Accelerator Center 152T
STC 126T, 139B
Sterling-Winthrop Group 61T
Swiss National Tourist Office 79L, 80B
D. T. Thomson/Science Photo Library title page
Tricity 36B
TVS 122, 132T
UKAEA 17, 18, 67B, 68, 142, 156B
Van den Berghs & Jergens 36T
Vickers Oceanics 63B
Westinghouse Electric 24B
Zeiss 156T